4th International Conference on Materials Engineering and Nano Sciences (ICMENS 2020)

IOP Conference Series: Materials Science and Engineering
Volume 840

Pattaya, Thailand
13 - 15 March 2020

ISBN: 978-1-7138-2682-8
ISSN: 1757-8981

Printed from e-media with permission by:

Curran Associates, Inc.
57 Morehouse Lane
Red Hook, NY 12571

Some format issues inherent in the e-media version may also appear in this print version.

Institute of Physics
Dirac House, Temple Back
Bristol BS1 6BE UK

Phone: 44 1 17 929 7481
Fax: 44 1 17 920 0979

techtracking@iop.org

Additional copies of this publication are available from:

Curran Associates, Inc.
57 Morehouse Lane
Red Hook, NY 12571 USA
Phone: 845-758-0400
Fax: 845-758-2633
Email: curran@proceedings.com
Web: www.proceedings.com

TABLE OF CONTENTS

NANOMATERIALS

STABILITY TEST OF NANOSTRUCTURED LIPID CARRIERS-LOADED MEFENAMIC ACID PREPARED BY MICROEMULSION TECHNIQUE 1
Jirapornchai Suksaeree, Apinya Treelop, Pitsacha Veeravatanayothin, Pattwat Maneewattanapinyo, Chaowalit Monton

PERFORMANCE OF MODIFIED NANO-SIO2 COMPOSITE PHOSPHATING COATING ON THE SURFACE OF STEEL 6
Xuxiang Tan, Feng Nan

A PRELIMINARY STUDY OF A GRAPHENE FRACTAL SIERPINSKI ANTENNA 11
Alberto Boretti, Lorenzo Rosa, Jonathan Blackledge, Stefania Castelletto

ELECTRONIC PROPERTIES OF [111] HYDROGEN PASSIVATED GE NANOWIRES WITH SURFACE SUBSTITUTIONAL LITHIUM 19
L G Arellano, F Salazar, A Trejo Baños, A Miranda, L A Pérez, M Cruz-Irisson

MATERIAL PROCESSING

DECOMPOSITION OF THE VECTOR FIELD OF PREFERRED DIRECTIONS FOR OPTIMIZATION OF FIVE-AXIS MACHINING 24
S. S. Makhanov

ACCURATE SCALLOP EVALUATION METHOD CONSIDERING KINEMATICS OF FIVE-AXIS MILLING MACHINE FOR BALL-END MILL 29
Luu Hai Tuan, Stanislav S Makhanov

MATERIAL PHYSICS

EFFECTS OF TRANSITION ZONE ON MAGNETIC PROPERTIES OF LOW TEMPERATURE OXIDATION OF MAGNETITE PARTICLE:COMPARISON OF EXPERIMENT AND MICROMAGNETIC MODELING 35
K Ge, W Williams

REFLECTION SPECTRA OF A THIN-FILM GESBTE DIFFRACTION GRATING ON A SILICON NITRIDE WAVEGUIDE 40
R T Minnullin, A A Sapegin, M E Makarov, D S Korolev, M Yu Barabanenkov

EFFECTS OF SURFACE IN THE IR AND RAMAN SPECTRUM OF POROUS SILICON CARBIDE 46
R Bermeo, L. Arellano, A Trejo, F Salazar, M. Calvino, A Miranda, M Cruz-Irisson

VISIBLE AND INFRARED PHOTOLUMINESCENCE IN HEXAGONAL SILICON CARBIDE BY DIRECT FEMTOSECOND LASER WRITING 51
Stefania Castelletto, Brett C Johnson, Alberto Boretti

MATERIAL CHARACTERIZATION

MECHANICAL AND THERMAL BEHAVIOUR OF EPOXY/HEXAGONAL BORON NITRIDE/SHORT SISAL FIBER HYBRID COMPOSITES 59
Alok Agrawal, Saurabh Chandrakar, Abhishek Sharma

MECHANICAL BEHAVIOR ANALYSIS OF STAINLESS STEELS SUBJECTED TO UNIAXIAL STRESS TESTS 66
J Brnic, S Krscanski, M Brcic

MORPHOLOGICAL AND MECHANICAL PROPERTIES OF NATURAL RUBBER COMPOUND/POLY(BUTYLENE SUCCINATE) BLEND 72
N Prasoetsopha, P Thainoi, R Jinnavat, W Charerntanom, A Hasook, W Singsang

STUDY ON THE INFLUENCE OF BLAST FURNACE SLAG ON CHLORIDE ION PENETRATION PROPERTY 77
P Subpa-asa, K Ken, F Satoshi, D Shigeyuki

MATERIALS FOR ENERGY APPLICATION

UTILIZATION OF WASTE LUBRICATING OIL AS A DIESEL ENGINE FUEL 83
Abhishek Sharma, Gaurav Gupta, Alok Agrawal

UTILISATION OF EGGSHELL POWDER (ESP) AS PARTIAL REPLACEMENT OF CEMENT INCORPORATING SUPERPLASTICIZER 93
Gunalaan Vasudevan, Seah Chan Wei

CASE STUDIES OF HIGH-STRENGTH LIGHTWEIGHT CONCRETE USING EXPANDED SILICEOUS AGGREGATE 100
A S Inozemtcev

Author Index

IOP Conf. Series: Materials Science and Engineering **840** (2020) 011001 doi:10.1088/1757-899X/840/1/011001

Preface

Materials Engineering and Nano Sciences is a key ingredient of current advanced technologies supporting modern society. The tremendous rapid growth of these fields makes them highly interdisciplinary and then inevitably demands their mutual interactions. Under these circumstances, the 2020 4th International Conference on Materials Engineering and Nano Sciences (ICMENS 2020) was organized as a series of successful international conferences held in Singapore (2017), Hong Kong (2018) and Japan (2019).

The main goal of ICMENS 2020, which will be held from 13 to 15 March 2020, in Pattaya, Thailand, is to address the latest original results in materials engineering and the nano sciences, including both theoretical advances and practical implementations, which are becoming more and more popular in industry and in our daily lives. The ICMENS 2020 will provide a premier interdisciplinary platform for scientists, researchers, industry leaders, engineers and educators throughout the world to present and discuss the most recent innovations, trends, concerns, as well as practical challenges encountered, and streamline solutions in the fields of materials engineering and nano sciences.

Submissions were received from 10 countries, including China, Japan, Russia, Czech Republic, Saudi Arabia, Thailand, Malaysia, Croatia, India, and Mexico. After several rounds of paper submission and reviewing, 17 papers have been selected to be published in the IOP conference proceedings.

On behalf of the Organising Committee, I wish to thank the keynote speakers, invited speakers and authors of the selected papers for their outstanding contributions. I would also like to thank members of the organizing committee, anonymous reviewers and volunteers for their great efforts. Without their contribution, dedication and commitment, we would not have achieved so much. Thanks also go to the Conference Secretaries (Dana Chan & Jessi Mars) for their dedicated cooperation. We sincerely hope that you will find the ICMENS 2020 beneficial and fruitful for your professional development. We also hope that you will enjoy our hospitality and will have an enjoyable and memorable time in Pattaya.

Conference Chair, ICMENS 2020

Prof. Richard Haverkamp
Massey University, New Zealand

IOP Conf. Series: Materials Science and Engineering **840** (2020) 011002 doi:10.1088/1757-899X/840/1/011002

Conference Committees

Advisory Committees
Prof. Akira Toriumi, IEEE Fellow, The University of Tokyo, Japan
Prof. Alan Lau, Swinburne University of Technology, Australia

Conference Chairs
Prof. Dr. -Ing. Thira Jearsiripongkul, Dean, Thammasat University, Thailand
Prof. Richard Haverkamp, Massey University, New Zealand

Local Chair
Prof. Cattaleeya Pattamaprom, Thammasat University, Thailand

Program Chairs
Prof. Moo Hwan Cho, Yeungnam University, South Korea
Prof. Ricardo Branco, University of Coimbra, Portugal
Prof. Akihiko FUJIWARA, Kwansei Gakuin University, Japan
Prof. Jui-Hsiang Liu, National Cheng Kung University (NCKU), Taiwan

Program Co-chairs
Prof. Syuma Yasuzuka, Hiroshima Institute of Technology, Japan
Assoc. Prof. Petr Valášek, Czech University of Life Science Prague, Czech Republic
Prof. Josip Brnić, University of Rijeka, Croatia

Technical Committees
Assoc. Prof. Bashir Ahmmad, Yamagata University, Japan
Dr. Kiattipoom Rodpun, Mahidol Wittayanusorn School, Thailand
Dr. Patittar Nakwong, Phranakhon si Ayutthaya Rajabhat University, Thailand
Dr. Gunalaan Vasudevan, Tunku Abdul Rahman University College, Malaysia
Assoc. Prof. Marino Brčić, University of Rijeka, Croatia
Dr. Francesca Scalisi, Euro-Mediterranean Documentation and Research Center, Italy
Prof. Rakesh Chand Ramola, HNB Garhwal University, India
Dr. Abbas Mohammed Selman, Universiti Sains Malaysia, Malaysia
Dr. Heah Cheng Yong, Unversiti Malaysia Perlis, Malaysia
Dr. Liew Yun Ming, Unversiti Malaysia Perlis, Malaysia
Dr. Trong-Phuoc Huynh, Can Tho University, Vietnam
Dr. Nur Farhana Diyana Mohd Yunos, Unversiti Malaysia Perlis, Malaysia
Dr. K. Krishnamoorthy, National Chemical Laboratory, India
Dr. Sujarinee Sinchai (Kochawattana), Chulalongkorn University, Thailand
Asst. Prof. Ibetombi Soibam, National Institute of Technology Manipur, India
Dr. Zhauyt Algazy, Kazakh National Research Technical University, Kazakhstan
Dr. K. Krishnamoorthy, Anna University, India

First name	Surname	Institution
Jirapornchai	Suksaeree	College of Pharmacy, Rangsit University, Thailand
Xuxiang	Tan	Shandong Jiaotong University, China
Alberto	Boretti	Department of Mechanical Engineering, College of Engineering, Prince Mohammad Bin Fahd University, Al Khobar, Saudi Arabia
Lucia Guadalupe	Arellano Sartorius	Instituto Politécnico Nacional, ESIME Culhuacán, Av. Santa Ana 1000, 04430 Ciudad de México, México
Stanislav S	Makhanov	Thammasat University, Thailand
Luu Hai	Tuan	Thammasat University, Thailand
Kunpeng	Ge	East China University of Technology, China
R T	Minnullin	JSC Molecular Electronics Research Institute, Moscow, Zelenograd, 124460, Russia and Moscow Institute of Physics and Technology, Dolgoprudny, 141701, Russia
Alejandro Trejo	Baños	Instituto Politécnico Nacional, ESIME Culhuacan, Av. Santa Ana 1000, C.P. 04440, Ciudad de México, Mexico, México
Stefania	Castelletto	School of Engineering, RMIT University, Bundoora, Victoria, Australia and Swinburne Nanotechnology facility, Centre for Micro-Photonics, Swinburne University of Technology, Hawthorn, Victoria, Australia
Alok	Agrawal	Sagar Institute of Research and Technology-Excellence, India
Josip	Brnic	University of Rijeka, Faculty of Engineering, Croatia
Natkrita	Prasoetsopha	Rajamangala University of Technology Isan, Thailand
Prang	Subpa-asa	Tokai University, Japan
Abhishek	Sharma	Department of Mechanical Engineering, Manipal University Jaipur, Rajasthan, India
Gunalaan	Vasudevan	Tunku Abdul Rahman University College, Malaysia
Aleksandr Sergeevich	Inozemtcev	National Research Moscow State University of Civil Engineering, Russia

E-mail address
jirapornchai.s@rsu.ac.th
tanxx1982@163.com
a.a.boretti@gmail.com
lucia.arellano.gin2017@gmail.com
makhanov@siit.tu.ac.th
xhaituanx@gmail.com
kunpeng.ge@yahoo.com
ramil.minnullin@phystech.edu
alejandtb13@gmail.com
stefania.castelletto@rmit.edu.au
alokag03@gmail.com
brnic@riteh.hr
natkrita.pr@rmuti.ac.th
prangsub@gmail.com
drasharma58@gmail.com
gunalaanv@tarc.edu.my
InozemcevAS@mgsu.ru

IOP Conf. Series: Materials Science and Engineering **840** (2020) 011004 doi:10.1088/1757-899X/840/1/011004

Peer review statement

All papers published in this volume of *IOP Conference Series: Materials Science and Engineering* have been peer reviewed through processes administered by the proceedings Editors. Reviews were conducted by expert referees to the professional and scientific standards expected of a proceedings journal published by IOP Publishing.

IOP Conf. Series: Materials Science and Engineering **840** (2020) 012001 doi:10.1088/1757-899X/840/1/012001

Stability Test of Nanostructured Lipid Carriers-Loaded Mefenamic Acid prepared by Microemulsion Technique

Jirapornchai Suksaeree[1], Apinya Treelop[1], Pitsacha Veeravatanayothin[1], Pattwat Maneewattanapinyo[1] and Chaowalit Monton[2]

[1] Department of Pharmaceutical Chemistry and [2] Drug and Herbal Product Research and Development Center, College of Pharmacy, Rangsit University, Muang, Pathum Thani 12000, Thailand

E-mail: jirapornchai.s@rsu.ac.th

Abstract. The objectives of this study were to prepare the nanostructured lipid carriers (NLCs)-loaded mefenamic acid and to study the particle size and zeta potential of the NLCs-loaded mefenamic acid after storage in refrigerator (approximately 2 – 4°C) for 7 days. The preparation method of NLCs-loaded mefenamic acid used the microemulsion technique. All compositions were vigorously mixed with vertex method to obtained the clear microemulsion solution. Tween®80 (surfactant) and 1 – butanol (co – surfactant) were mixed at the ratio of 3:2 to use as the surfactant mixture. Then, it was dispersed in cold water (approximately 2 – 4°C) at different ratio, and the droplet size was reduced by a homogenizer at 6,000 rpm for 5 minutes and sonicator for 30 minutes to form the NLCs-loaded mefenamic acid. The prepared NLCs-loaded mefenamic acid were kept in refrigerator (approximately 2 – 4°C) for 7 days. After that, the solutions were tested the particle size and zeta potential. The particle size and zeta potential of NLCs-loaded mefenamic acid after storage were in the range of 160-310 nm and -4.00 to -19.00 mV, respectively. It was found that the F4 formula was the best to prepare the NLCs-loaded mefenamic acid. F4 formula composed of the 1%w/w mefenamic acid, 10%w/w water, 80%w/w surfactant, and 10%w/w capric triglyceride, and the dilution ratio was 1:25.

1. Introduction

Nanostructured lipid carriers (NLCs) are the lipid nanoparticles that produce by blending of a solid lipid and a liquid lipid, this blend also being solid at body temperature. The solid lipid and liquid lipid blend are melted and hot aqueous surfactant or stabilizer solution is added. The obtained pre – emulsion is produced to form the NLCs by various preparation techniques such as hot and cold high-pressure homogenization, high shear homogenization and ultrasound, microemulsion technique, and solvent emulsification/evaporation process. When the solution is cool down, the emulsion droplets crystallize forming lipid nanoparticles with solid particle matrix [1-3]. The advantage of the NLCs is the increased loading of active compound compared to solid lipid nanoparticles and firmer inclusion of the active compound inside the particle matrix during the shelf life. The use of a lipid mixture between solid lipid and a liquid lipid distorts the formation of very differently structured molecules that present a perfect crystal with high space for accommodate the active compound in molecular form or as amorphous clusters [3].

Mefenamic acid is classified in a Biopharmaceutics Classification System II, low solubility and high permeability drug. which can dissolve in water at 0.21 mg/mL; in addition, it can the solubility is greatly increased when co-solvent is added such as ethanol, propylene glycol, ethylene glycol,

IOP Conf. Series: Materials Science and Engineering **840** (2020) 012001 doi:10.1088/1757-899X/840/1/012001

polyethyleneglycol-400, glycerin, and 10% surfactant solutions (Tween®80, sodium lauryl sulfate, or polyethylene glycol dodecyl ether). Ethanol and polyethyleneglycol-400 are reported to be the most effective to enhance the solubility to more than 11 mg/mL [4]. Mefenamic acid is a non-steroidal anti-inflammatory drug indicated for relief of mild-to-moderate pain and primary dysmenorrhea by inhibition of cyclo-oxygenase-2 and prostaglandin synthesis. It is available in the as tablets, capsules, and pediatric suspensions. It has a short elimination half–life which 2 hours. Thus, oral administration requires frequent dosing every 6 hours in order to maintain the steady-state plasma concentration. Recent effort to increase the drug level in plasma more than 2 hours employed controlled release beads [5], sustained release matrix tablets [6, 7], and controlled release tablets [8].

However, crystallization of mefenamic acid has been found after preparing it for use in transdermal patches [9]. Thus, the aim of this study was to prepare the NLCs-loaded mefenamic acid from microemulsion method. The clear microemulsion was prepared and was dispersed in cold water and its droplet size was reduced by homogenizer and sonicator to form the NLCs. Capric triglyceride was used as a liquid lipid phase. The distilled water was used as water phase. Tween®80 and 1-butanol were used as a surfactant and a co-surfactant, respectively. The characterization of NLCs-loaded mefenamic acid were analysed after storage in refrigerator (approximately 2 – 4°C) for 7 days.

2. Experimental works

2.1 Materials

Mefenamic acid and 1-Butanol were obtained from Sigma-Aldrich (USA). Capric triglyceride and Tween®80 were obtained from P.C. Drug Center Co. ltd (Thailand).

2.2 Preparation of NLCs-loaded mefenamic acid prepared by microemulsion technique

The compositions of NLCs-loaded mefenamic acid are shown in Table 1. Briefly, the water, capric triglyceride, and surfactant, were vigorously mixed with vertex method to obtained the clear microemulsion solution. Then, it was dispersed in cold water (approximately 2 – 4°C) at different ratio, and the droplet size was reduced using a homogenizer at 6,000 rpm for 5 minutes and a sonicator for 30 minutes to form the NLCs-loaded mefenamic acid. Tween®80 (surfactant) and 1 – butanol (co – surfactant) were mixed at the ratio of 3:2 to obtain the mixture surfactant. The stability testing of NLCs-loaded mefenamic acid, the prepared NLCs-loaded mefenamic acid were kept in refrigerator (approximately 2 – 4°C) for 7 days. After that, the solutions were tested the particle size and zeta potential.

Table 1. The formulas of NLC-loaded mefenamic acid prepared by microemulsion technique.

Formulas	Mefenamic acid	Water	Surfactant (Tween®80:1-Butanol=3:2)	Capric triglyceride	Dilution ratio
F1	1%w/w	10%w/w	70%w/w	20%w/w	1:25
F2	1%w/w	10%w/w	70%w/w	20%w/w	1:50
F3	1%w/w	10%w/w	70%w/w	20%w/w	1:100
F4	1%w/w	10%w/w	80%w/w	10%w/w	1:25
F5	1%w/w	10%w/w	80%w/w	10%w/w	1:50
F6	1%w/w	10%w/w	80%w/w	10%w/w	1:100

Table 2. Particle sizes, polydisperse index, and zeta potential of NLCs-loaded mefenamic acid.

Formulas	Particle size (nm)	Polydisperse index	Zeta potential (mV)
F1	201.1±18.5	0.358±0.024	-13.85±1.85
F2	191.3±23.0	0.364±0.046	-18.05±1.23
F3	304.5±71.2	0.336±0.052	-12.75±0.59
F4	216.9±48.8	0.291±0.042	-8.30±0.56
F5	306.7±105.1	0.279±0.048	-9.12±0.83
F6	160.8±23.9	0.563±0.028	-4.24±0.70

IOP Conf. Series: Materials Science and Engineering **840** (2020) 012001 doi:10.1088/1757-899X/840/1/012001

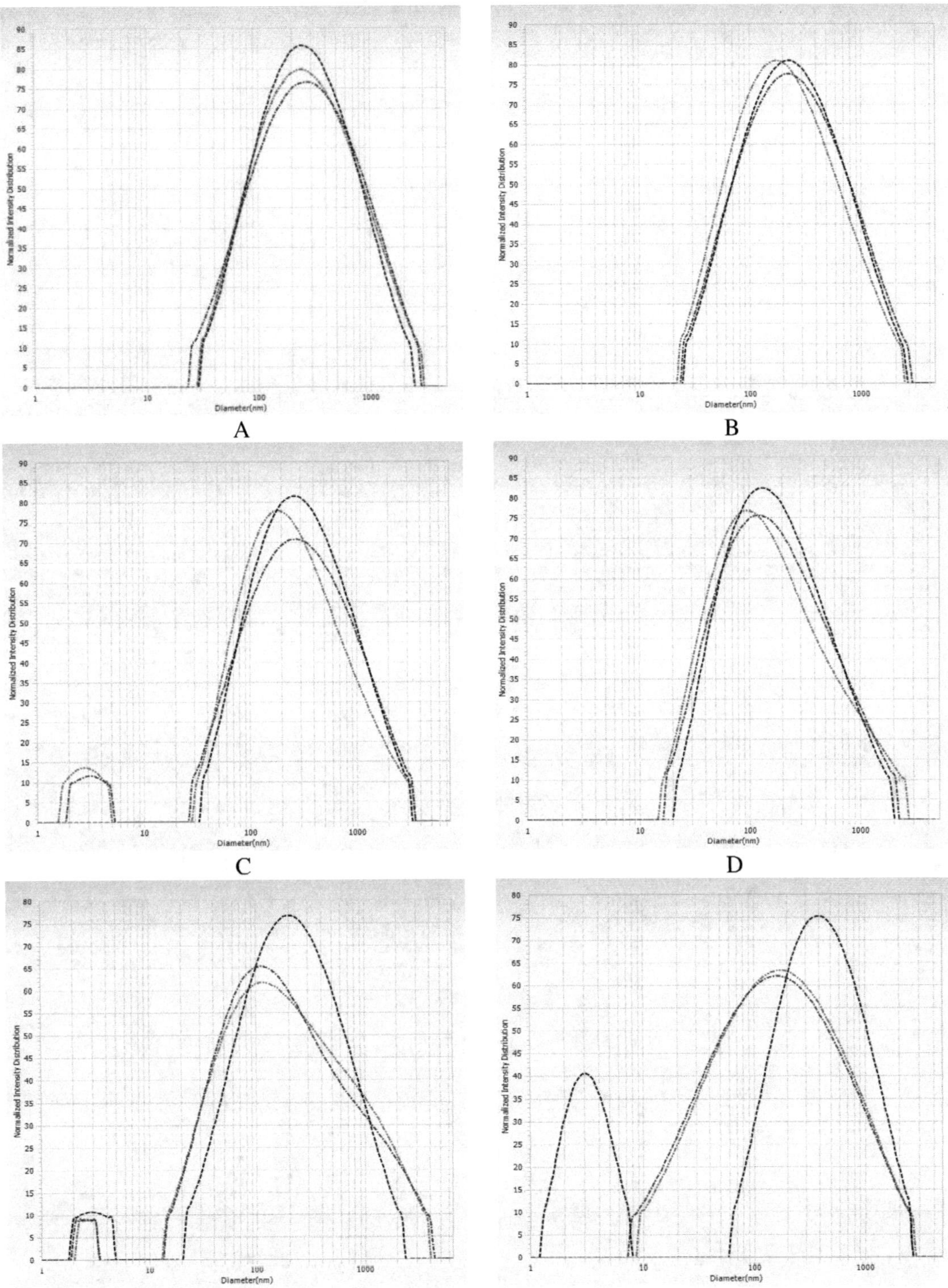

Figure 1. Particle distribution of the NLCs-loaded mefenamic acid prepared by microemulsion technique; F1 (A), F2 (B), F3(C), F4 (D), F5 (E), and F6 (F)

IOP Conf. Series: Materials Science and Engineering **840** (2020) 012001 doi:10.1088/1757-899X/840/1/012001

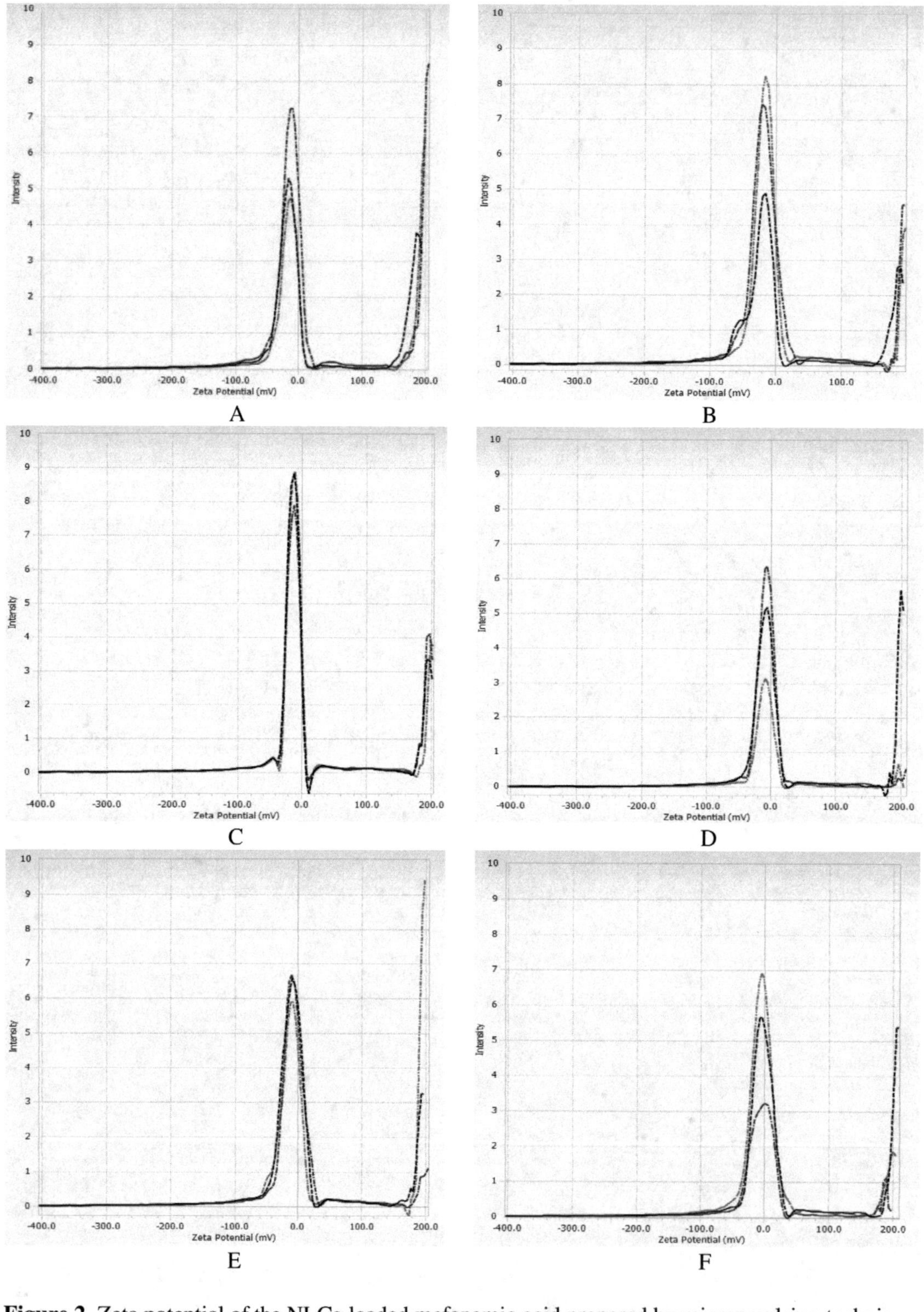

Figure 2. Zeta potential of the NLCs-loaded mefenamic acid prepared by microemulsion technique; F1 (A), F2 (B), F3(C), F4 (D), F5 (E), and F6 (F)

2.3 Particle size and zeta potential analysis

The distilled water was used as a dilutant at an appropriate concentration prior to NLCs-loaded mefenamic acid determination. The solution of prepared NLC-loaded mefenamic acid was diluted in 5 times. The Nanoplus 3 (Particulate system, USA) was used to determine the particle size, size distribution, and zeta potential of the NLCs-loaded mefenamic acid at 25±2ºC. The testing was recorded as a mean±SD for ten sub-runs.

3. Results and discussion

The distribution of particle size and zeta potential of NLCs-loaded mefenamic acid prepared by microemulsion technique after storage in refrigerator (approximately 2 – 4°C) for 7 days are shown in Figure 1 and 2, respectively. These values are concluded that in Table 2. It was found that the particle size was in the range of 160-310 nm with polydisperse index less than 0.6. After storage in refrigerator (approximately 2 – 4°C) for 7 days, all formulas had the high distribution of particle size. Their particles might be accumulated in the solution; however, the size of particles showed less than 500 nm, it was accepted to requirement of nanoparticle preparation. The F4 formula showed low distribution of particle size compared to other formulas. The zeta potential was in the range of -18 to -4 mV. When the amount of surfactant increased, the zeta potential of the formulation was increased. The best formula of NLCs-loaded mefenamic acid prepared by microemulsion technique after storage in refrigerator (approximately 2 – 4°C) for 7 days was F4 due to low particle size (216.9±48.8 nm), low polydisperse index (0.291±0.042), and the zeta potential as -8.30±0.56 mV. Therefore, F4 formulas will be improve and develop in the future to increase the stability.

4. Conclusions

The NLCs-loaded mefenamic acid was prepared from the microemulsion technique using the water as liquid phase, capric triglyceride as the oil phase, and mixture surfactant between Tween®80 and 1 – butanol. Its droplet size was reduced by homogenizer and sonicator after the microemulsion solution was dispersed in cold water. The prepared NLCs-loaded mefenamic acid were kept in refrigerator (approximately 2 – 4°C) for 7 days to study the stability. The particle size and zeta potential were shown in the range of 160-310 nm and -4.00 to -19.00 mV, respectively. The F4 formula was the best to prepare the NLCs-loaded mefenamic acid. F4 formula which composed of the 1%w/w mefenamic acid, 10%w/w water, 80%w/w surfactant, and 10%w/w capric triglyceride, and the dilution ratio was 1:25. Therefore, F4 formulas will be improve and develop in the future to increase the stability and it will be characterized and studied the *in vitro* release.

5. References

[1] Mehnert W and Mäder K 2001 *Adv. Drug Deliv. Rev.* **47** 165-196.
[2] Bunjes H 2011 *Curr. Opin. Colloid Interface* **16** 405-411.
[3] Müller RH, Petersen RD, Hommoss A and Pardeike J 2007 *Adv. Drug Deliv. Rev.* **59** 522-530.
[4] Swathi CH, Subrahmanyam S, Kedarnath P and Babu S 2011 *Int. J. Pharm. Tech.* **3** 267–276.
[5] Sevgi F, Kaynarsoy B and Ertan G 2008 *Pharm. Dev. Technol.* **13** 5-13.
[6] Güngör S, Yıldız A, Özsoy Y, Cevher E and Araman A 2003 *Il Farmaco* **58** 397-401.
[7] Adam A, Schrimpl L and Schmidt PC 2000 *Drug Dev. Ind. Pharm.* **26** 489-497.
[8] Patel P, Madan P and Lin S 2014 *Pharm. Dev. Technol.* **19** 355-362.
[9] Suksaeree J, Piamsap K, Paktham S, Kenprom T, Monton C and Pichayakorn W 2017 *Monatsh. Chem. Chem. Mon.* **148** 1215-1222.

Acknowledgment

The financial supporting of this work was the Undergraduate Scholarships from College of Pharmacy, Rangsit University.

IOP Conf. Series: Materials Science and Engineering **840** (2020) 012002 doi:10.1088/1757-899X/840/1/012002

Performance of Modified Nano-SiO2 Composite Phosphating Coating on the Surface of Steel

Xuxiang Tan[1], Feng Nan[1]

[1]School of Civil Engineering, Shandong Jiaotong University, 5001 Haitang Road, Changqing District, Jinan, Shandong Province, China

Email: tanxx1982@163.com

Abstract: In order to enhance the corrosion resistance and the quality of the phosphating film, nano-SiO2 was used in the phosphating solution.The structure and properties of the phosphating film were evaluated by SEM, electrochemistry, cupric sulfate drop corrosion test and film thickness measurement.The results show that the thickness of the phosphating film increases with the increase of nano-SiO2 content. The nano-SiO2 modified with vinyl triethoxy silane possessed good dispersibility in the phosphating solution, the phosphating solution with the addition of modified nano-SiO2 could greatly improve the corrosion resistance of the phosphating film.

1. Introduction

Phosphating is almost a standardized pretreatment mode for cathodic electrophoretic coating. Phosphating can enhance the corrosion resistance of metal, provide more fixing points for paint film and increase the adhesion of paint film[1].Most phosphating solutions contain nickel ions. Adding nickel ion can not only speed up phosphating treatment, but also improve film compactness, adhesion and corrosion resistance[2-3].However, nickel is a heavy metal ion, which is harmful to human body and environment.The emission standard of nickel ion has been reduced from 1.0 mg / L to 0.5 mg / L due to increasingly strict restrictions[4].It is still an important research direction of phosphating industry in the future to explore the phosphating process with little or no nickel.In recent years, with the development of nanotechnology, nano materials are widely used in traditional surface treatment fields, such as solid coating, organic coating, passivation[5-7].However, there are relatively few applications of nanomaterials in phosphating processes, and related research is still in its infancy.For example, R. Mohammad Hosseini et al. [8] studied the phosphating solution added with nano-Al_2O_3 and found that the phosphating film has strong wear resistance and lubricity.M. Tamilselvi et al. [9]prepared phosphate coatings on the surface of a steel plate, and found that the phosphate coating containing nano ZnO had better corrosion resistance.S.M.A.Shibli[10]reviewed the film-forming process of phosphating solution containing nano-TiO2 on the surface of galvanized steel plate. It was noted that the addition of nano-TiO2 could improve the electrochemical stability and corrosion resistance of the film.

In this study, the nano-SiO_2 was added into the phosphating solution instead of nickel ion. A phosphating solution containing no nickel ions was prepared.

2. Materials and experiments

2.1 Nano-SiO_2 preparation
Add 50ml of sodium silicate solution of a certain concentration to the three mouth flask, put it on the mixer, adjust the rotating speed to 100 r/ min, and slowly raise the temperature to 60 °C, put the flask into a constant temperature water bath at 60 °C, slowly add 1mol / L sulfuric acid to the flask, and quickly stir it to make it mix evenly until white precipitate appears.

Weigh 0.3 g of nano-SiO_2 and dissolve it in ethanol solution, stir it at room temperature and disperse it in ultrasonic cleaner for 15 min to obtain uniform and transparent suspension.In addition, 0.3g vinyl triethoxysilane (A151) was hydrolyzed in 3g deionized water at room temperature under acidic conditions, and pH = 3-4 was adjusted with phosphoric acid until the hydrolysis was complete.

2.2 Phosphating process
The mild steel is first sanded with 60# sandpaper to remove the surface deposits, and then the surface of the steel bar is completely exposed by grinding with 120# sandpaper. After polishing, rinse with water and rinse for 1 min;Rinse the steel bar with a dilute hydrochloric acid solution with a mass concentration of 5% to activate the steel bar metal for 30 s; then rinse with water for 30 s to remove residual acid on the surface;The steel bar is placed in the phosphating solution for 5 to 10 minutes; the cleaning is carried out for 1 min to remove the residual phosphating solution and dried at room temperature. The above test procedures were all carried out at room temperature.

2.3 Preparation of phosphating solution
The basic test phosphating formula is obtained through preliminary orthogonal test, as shown in Table 1.

Table 1. Formula of phosphating solution g • L-1

composition	ZnO	H_3PO_4	$Mn(NO_3)_2 \cdot 4H_2O$	$NaNO_3$	NaF	citric acid
mixture ratio	1.5	18.0	9	16	0.8	1.2

The composite modified nano-SiO2 phosphating solution was prepared by adding the modified nano-SiO2 dispersion, stirring and dispersing, and then ultrasonic treatment for 15 min.After the phosphating solution is prepared, adjust its acidity to the process range with 1 mol / L NaOH solution.

2.4 Instruments and analytical methods

2.4.1 Morphology analysis. The morphology of phosphating film was observed by JSM-6700F scanning electron microscope (jeol company of Japan)

The thickness of phosphating film was measured by thickness gauge. Take three points of the same sample for measurement, and calculate the arithmetic mean value, which is the thickness of the phosphating film.

2.4.2 Cupric sulfate drop corrosion test. Take a drop of copper sulfate solution (40ml 0.1mol/L copper sulfate solution + 20ml 10% sodium chloride solution + 0.8ml 0.1mol/L hydrochloric acid) and drop it on the phosphating film. Observe the time from blue, light green, yellow to red. Take more than three points to calculate the average value.

2.4.3 Electrochemical characteristics of steel fiber phosphating film formation process. The test material is a steel fiber sample. The sample for the Et (potential-time) curve is made into a working electrode of 1 cm×0.1 cm, the electrolyte is a phosphating solution, and the electrochemical test uses a three-electrode system. The specific electrode is a saturated calomel electrode, the auxiliary electrode is a platinum wire, and the test instrument adopts a CP-6 type potentiostat. The temperature of the phosphating solution is maintained at 20±1°Cin a constant temperature water bath during the test.

IOP Conf. Series: Materials Science and Engineering **840** (2020) 012002 doi:10.1088/1757-899X/840/1/012002

2.4.4 Neutral salt spray test (NSST) corrosion. The corrosion solution is 50g / L NaCl solution, the solution Ph = (7+0.5), the temperature inside the salt spray box is (35+2)°C, the settlement is 2ml/ (80cm2·h), the sample is placed at 30 degrees in the vertical direction of the salt mist rack, and is continuously sprayed with 8h for 1 cycles.The corrosion area was tested and the corrosion resistance of phosphating film was evaluated after different cycles.

3. Results and discussion

3.1 Scanning electron micrograph

Through the analysis of the surface morphology after phosphating and corrosion (1500 times), it can be found that the appearance of the phosphated steel fiber is dark gray, continuous and uniform, as shown in Figure 1. It can be seen from Figure 1that the generated phosphating film completely covers the metal substrate and is continuously distributed in blocks with large surface area and completely covers the metal surface.

3.2 Effect of nano-SiO_2 on phosphating film thickness

It can be seen from Figure 2 that the film thickness of the phosphating film increases with the increase of the concentration of nano-SiO2.

3.3 Effect of Nano-SiO_2 on corrosion resistance of phosphating film

It can be seen from Figure 2 that the corrosion resistance time of copper sulfate in phosphating solution increases first and then decreases with the increase of nano-SiO2 content. When the content of nano-SiO2 is 1g/L, the corrosion resistance time of the phosphating film is the best, more than 115s.

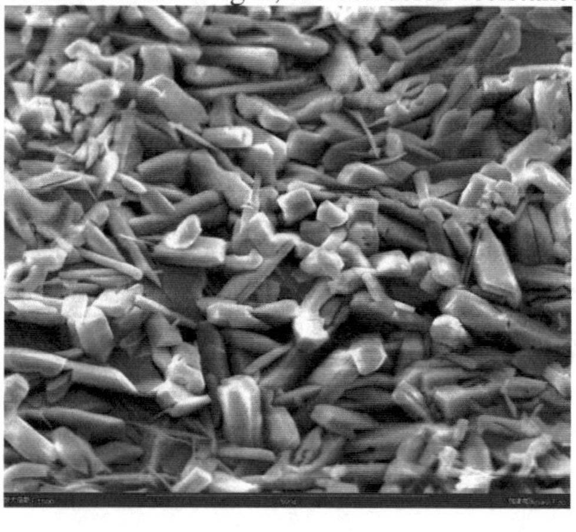

Figure 1. phosphating film SEM

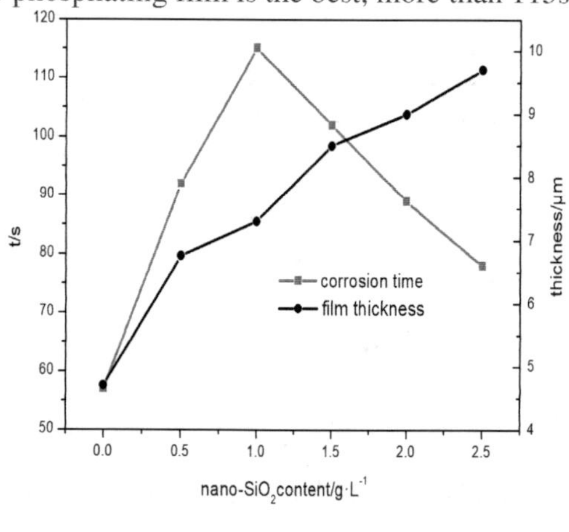

Figure 2. Effect of nano-SiO_2 content on film thickness and Cupric sulfate drop

3.4 Electrochemical analysis

In order to simulate the corrosion resistance of phosphated steel, unphosphorized steel and stainless steel in the corrosive environment, adjust the mass fraction of sodium chloride in the solution to 3.5%, and the measured potential changes are shown in Figure 3.It can be seen from Figure 3 that the potential of the modified phosphating film is the highest, and its rise range with the change of time potential is the smallest.

3.5 Neutral salt spray performance

Figure 4 shows the corrosion area curve of steel before and after phosphating in different neutral salt spray cycle tests.It can be seen from Figure 4 that the corrosion area of unphosphorized steel is about 80% after the second salt spray cycle, and it is completely corroded after the third salt spray cycle. However, the corrosion area of galvanized steel plate after phosphating treatment is significantly smaller than that of unphosphorized steel plate.After five cycles of salt spray, the corrosion area of the composite modified nano-SiO2 phosphating film is about 38%, while the corrosion area of the basic phosphating film is about 60%, which shows that the corrosion resistance of galvanized steel after the composite modified nano-SiO2 phosphating treatment is greatly improved on the basis of the basic phosphating treatment.

3.6 Mechanism analysis

As the nano-SiO2 still has the characteristics of nano particles, the specific surface area is large, the surface activity is high, and it is easy to adsorb on the steel surface.In the process of phosphating, when the galvanized steel is immersed in the phosphating solution, these particles will be adsorbed on the surface of steel, which not only increases the number of micro cathodes on the surface of the substrate, but also becomes a better crystal nucleus when the phosphate crystal is crystallized.In the process of film formation, such a large number of nuclei can change the extension direction of phosphate crystal crystallization, and make the crystal grow along its extension direction.With the progress of phosphating reaction, when the crystallization and dissolution of phosphate crystal reach equilibrium, these modified nano-SiO2 particles can be adsorbed on the surface of the film and filled in the gap, which makes the porosity of the phosphating film reduce and the corrosion resistance increase.

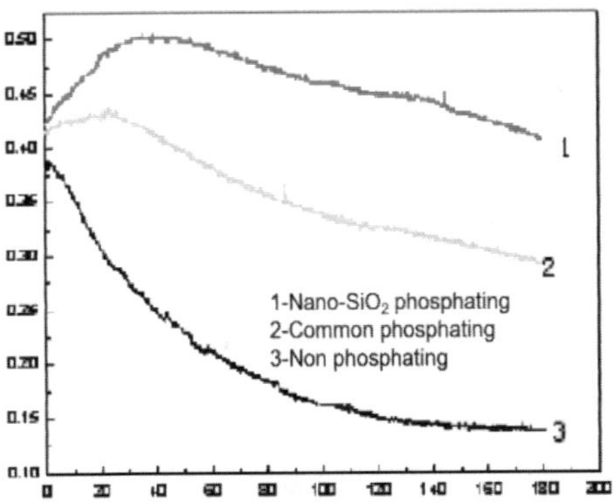

Figure 3. Potential change diagram in NaCl solution with different treatment methods

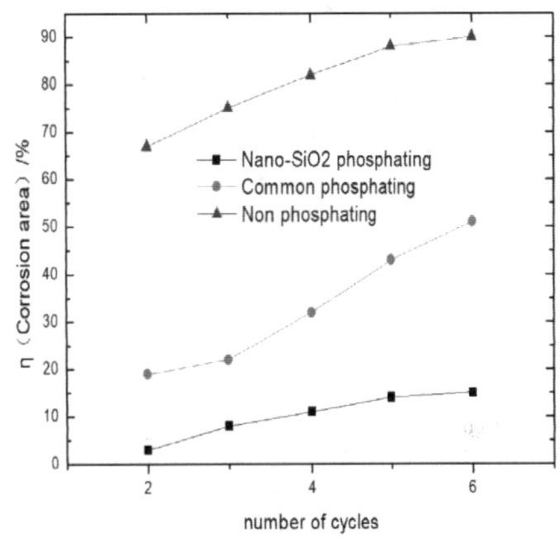

Figure 4. corrosion area curvein in different neutral salt spray cycle tests

4. Conclusion

The thickness of the phosphating film increases with the increase of the concentration of nano-SiO_2. Adding nano-SiO_2 to the phosphating solution can enhance the corrosion resistance of the phosphating film. When the amount of nano-SiO_2 is 1g / L, the thickness of the film is 7.81 μm, and the corrosion resistance of the phosphating film is the best.

The composite modified nano-SiO_2 phosphating film on the steel surface has excellent corrosion resistance, which can effectively prevent the corrosion of the film by the medium.

5. References

[1] J.B. Bajat,V.B. Mišković-Stanković,J.P. Popić,D.M. Dražić 2008 *Prog. Org. Coat.* **63** 201
[2] Yangshuhan Xu,Minghua Li,Mingyan Liu 2019 *Prog. Org. Coat.* **134** 177
[3] M. Fouladi, A. Amadeh 2013 *Mater. Lett.* **98** 1
[4] O. Lavigne,C. Alemany-Dumont,B. Normand,M.H. Berger 2011 *Corrosion Sci.* **53** 2087
[5] Thomas Lostak, Artjom Maljusch, Björn Klink, Stefan Krebs, Matthias Kimpel, Jörg Flock, Stephan Schulz, Wolfgang Schuhmann 2014 *Electrochim. Acta* **137** 65

[6] Xiuzhi Zhang,Fuhui Wang,Yuanlong Du 2007 *Surf. Coat. Technol.* **201** 7241
[7] Sviatlana V. Lamaka,Mikhail L. Zheludkevich,Kiril A. Yasakau,M 2005 *Electrochem. Commun.* **8** 421
[8] R. Mohammad Hosseini, A.A. Sarabi, H. Eivaz Mohammadloo, M. Sarayloo 2016 *Surf. Coat. Technol.* **206** 4132
[9] Fath, L.,Azadbeh, M. 2013 *World Appl. Sci.* **26** 122
[10] M. Tamilselvi,P. Kamaraj,M. Arthanareeswari,S. Devikala 2015 *Appl.Surf. Sci.***327** 218

Acknowledgments
This work was supported by Shandong Provincial Natural Science Foundation (ZR2017LEE029).

A Preliminary Study of a Graphene Fractal Sierpinski Antenna

Alberto Boretti[1,*], Lorenzo Rosa[2,3], Jonathan Blackledge[4,5,6], Stefania Castelletto[7]

[1]Department of Mechanical Engineering, College of Engineering, Prince Mohammad Bin Fahd University, Al Khobar, Saudi Arabia
[2]Department of Engineering "Enzo Ferrari", the University of Modena and Reggio Emilia, Modena, Italy
[3]Applied Plasmonics Lab, Centre for Micro-Photonics, Swinburne University of Technology, Hawthorn, Victoria, Australia
[4]School of Electrical and Electronic Engineering, Technological University Dublin, Ireland
[5]Faculty of Science and Technology, University of Wales, Wrexham, United Kingdom
[6]Department of Computer Science, University of Western Cape, Cape Town, South Africa
[7]School of Engineering, RMIT University, Bundoora, Victoria, Australia

*Email: a.a.boretti@gmail.com

Abstract. We provide a preliminary study of a Graphene fractal antenna operating at THz frequencies with the opportunity to modulate the emission. There are many advantages of the fractal design, namely multiband/wideband ability, and, a smaller, lighter and simpler configuration for higher gain, that can benefit from the coupling with Graphene, the thinnest and strongest of materials exhibiting very high electrical conductivity and tunability. This paper proposes a conceptual background for the study and presents some preliminary results on the electromagnetic emission simulations undertaken.

1. Introduction

Fractal antennas are receiving attention for the future of wireless communication because of their wideband and multiband capabilities, the opportunity of fractal geometries to drive multiple resonances and the ability to make smaller and lighter antennas, with fewer components and radiative elements less circuitry and higher gains. Extremely small and extremely high-frequency nanometric fractal antennas based on Graphene, a one-atom-thick, two-dimensional carbon crystal, may enhance wireless communications for commercial and military applications.

Nanoantennas based on surface plasmon polaritons enable the conversion of light from free space into sub-wavelength volumes establishing a way of communication using free-electron propagation within networks of nanosized devices. This approach can be of high impact for many applications, including biochemical sensors, reconfigurable meta-surfaces, compact optoelectronic devices, advanced health monitoring, drug delivery systems and wireless nano-sensor networks for biological and chemical attack prevention. Dynamic control and reconfigurable properties of these antennas are also very desirable for the above applications. Owing to its unique electronic properties, Graphene has recently been identified as a promising platform to build integrated active plasmonic nanoantennas for a wide wavelength range in the mid-infrared, i.e. 10 - 100 THz.

A Graphene fractal antenna is regarded as a high-frequency tunable antenna for radio communications in the Terahertz bandwidth, enabling unique applications in the military field, such as wireless nanosensor networks for biological and chemical attack prevention. Graphene is a one-atom-

IOP Conf. Series: Materials Science and Engineering **840** (2020) 012003 doi:10.1088/1757-899X/840/1/012003

thick two-dimensional allotrope of Carbon [1], with the highest known electrical conductivity that is currently unavailable in any other materials, including metals such as gold and silver. The nanostructures of Graphene can enable wireless communication enhancements within nanoscale electronic devices with a significant reduction in the energy needed for such communications [2]. Furthermore, as the electrons oscillate, they create an electromagnetic wave on top of the Graphene layer, referred to as the Surface Plasmon Polariton (SPP) wave [3], [4]. This enables the antenna to operate at the lower end of the Terahertz frequency spectrum [5], [6], which can be more efficient than current copper-based antennas [7].

In this paper, we study a Sierpinski fractal triangle monopole antenna made of gold, graphene, silicon, and silicon dioxide by solving Maxwell's equation for a complex refractive index in the frequency domain over the range 10 - 700 THz [8]. A few layers of graphene are deposited on a nanometric layer of silica on a Si substrate. Every material in the nanostructure is modeled through the complex refractive index with real, n, and, imaginary parts, k. The values of n and k are obtained from experimental data and theoretical models. The design permits the application of a gate voltage to modify the graphene optical conductivity over the operating region of 10 - 300 THz. We compare this design in Graphene and the operation with different gate voltages to the same design in gold, identifying the differences of using Graphene under the influence of the gate voltage.

The present dimensions follow the computational constraints set up by the thickness of the fractal that cannot be thicker than 3 nm. A better design for the fractal antenna is to scale it up by an order of magnitude or more while retaining the same thickness. This can be investigated as soon as such a design is feasible. Such a design will be able to scan with an order of magnitude smaller frequencies (10 times larger wavelengths). In this work, we model the material properties for wavelengths of 30 - 0.4 μm (10 -750 THz).

Our approach differs from similar work in [9], for example, where Graphene is placed underneath a Copper patch and the fractal is on the 100 x 100 μm scale. Here, we consider a design that extends the frequency range of operation of a fractal antenna by introducing nanofeatures. Further, Graphene nanoantenna plasmonic properties [3], [6], [10] provide tunability in the Terahertz domain due to the Graphene surface conductivity, that can be modified by applying an electrostatic bias, which changes the chemical potential (Fermi level). In [10], Graphene is used as an electrically tunable load in a nanoscale metal antenna gap. This permits an in-situ control of the antenna's frequency. The Graphene patch antenna resonance frequency changes with its width and Sierpiński-type fractals are of growing interest because of their broadband capabilities for applications using plasmonic metallic materials which generate emissions in the optical frequency (40 - 750 THz), [11].

In this paper, the combination of Graphene plasmonics and Sierpiński-type fractals is studied for the first time for a broad-band Terahertz antenna with a focus on nanoelectronic wireless communications, [3]. We note that the combination of metallic structures and Graphene, which has gate-voltage dependent optical properties, has already provided an antenna with narrower frequency operations, tuneable between 35 and 75 THz, [10].

We aim to design a compact structure while pushing the order of the Sierpiński fractal to a high level by engineering nanometric details in Graphene. This fabrication can be achieved by Electron Beam Lithography patterning or 3D nano-patterning using focused ion beam lithography, as described in [6] and [11]. Graphene is transferred on a SiO_2 substrate to control the size and features at the nanoscale. The nanoantenna is based on the Sierpiński gasket fractal, with equilateral triangles as building blocks.

2. Materials and methods

We simulate the optical properties of Graphene nanoantennas and the electric field intensity distribution at different resonance frequencies in the mid-infrared and the visible range. This is based on using the well-established Graphene optical conductivity based on a Drude-like model known as Kubo's formula [12], [13], within the random phase approximation in the local limit adopted by [10]. Other quantum mechanical models have recently been introduced to describe Graphene surface conductivity [14]. However, based on the recent agreement between the semi-classical model and experimental results as given in [10], we argue that our model is a good represents of the proposed design. Additionally, we model the refractive index of Graphene in the visible range by using the experimental data from [15].

The optical or surface conductivity of Graphene is based on a semi-classical model of conductivity. The surface conductivity for a single layer of Graphene has been determined by [13] and is given by

$$\boldsymbol{\sigma_s = \sigma_{s,intra} + \sigma_{s,inter}} \quad (1)$$

where

$$\boldsymbol{\sigma_{s,intra}} = \frac{je^2 k_B T}{\pi\hbar^2(\omega+j2\Gamma)}\left[\frac{\mu_C}{k_B T} + 2\ln\left(e^{-\frac{\mu_C}{k_B T}} + 1\right)\right] \quad (2)$$

$$\boldsymbol{\sigma_{s,inter}} = \frac{je^2}{4\pi\hbar}\ln\left[\frac{2|\mu_C|-(\omega+2j\Gamma)}{2|\mu_C|+(\omega+2j\Gamma)}\right],\ \boldsymbol{k_B T \ll \mu_C} \quad (3)$$

In the above equations, μ_C is the chemical potential (Fermi energy), which depends on the charge carrier concentration, Γ is the scattering rate with $\tau=1/2\Gamma$ being the carrier relaxation lifetime, k_B is the Boltzmann constant, T is the temperature and ω is the angular frequency. The first term of Eq. (1) is attributed to intra-band transitions and the second term to inter-band transitions. When the Fermi level is below half of the photon energy ($E_{ph} = \hbar\omega$), the contribution from the inter-band transition (visible and UV domain) dominates the optical conductivity. The inter-band surface conductivity can be calculated numerically or given in the analytic form under certain approximations. Once the Fermi level is increased above half of the photon energy, interband transitions are diminished due to Pauli blocking and intra-band transitions (in the IR and mid-IR) play a dominant role. To generate the surface conductivity of Graphene, the intra-band conductivity term can be obtained analytically while the inter-band conductivity term needs to be determined numerically.

A Drude-like approximation for $\boldsymbol{\sigma_s}$, is valid for mid-infrared wavelengths, where the intra-band conductivity term usually dominates over the inter-band term; thus, an approximation can be obtained from Eq. (2) for 0.01 to 100 THz given by

$$\boldsymbol{\sigma_{s,intra}} \approx \frac{je^2\mu_C}{\pi\hbar^2(\omega+j2\Gamma)} \quad (4)$$

where the carriers relaxation time is $\boldsymbol{\tau^{-1}} = \frac{ev_f^2}{\mu\mu_C}$, μ being the carrier mobility which is typically 2,000 $cm^2/(V\ s)$ and $\boldsymbol{v_f}$, the Fermi velocity of Dirac fermions in Graphene is given by $\boldsymbol{v_f} \approx 8.5 \times 10^5\ ms^{-1}$ to 3×10^6. With a silica substrate, a measured value can be found which is 1.8×10^6, [16].

A specific conductivity model was derived in [3], by taking into account the impact of lateral electron confinement in Graphene nanoribbons, but the above models converge for Graphene strips which are 50 nm wide or more. Further, experimentally measured values of the surface conductivity of Graphene nanoribbons can be utilized to better model small nanofeatures [17]. A closed empirical expression [18] that resolves numerical singularities in the Graphene surface conductivity occurring at $\boldsymbol{\hbar\omega = 2\ \mu_C}$ can also be used for λ > 250 nm up to a few microns; this model provides comparable results to the model adopted here, at least in the visible spectrum.

In this paper, we consider the Graphene sheet optical conductivity within the Random-Phase Approximation (RPA) in the local limit [12], [19] and also adopted in [10]. This model has been validated against direct measurements of the Graphene bulk conductivity. According to this model, the surface optical conductivity is given by

$$\boldsymbol{\sigma_s(\omega)} = \frac{2je^2k_BT}{\pi\hbar^2(\omega+2j\Gamma)}\ln\left[2\cosh\frac{\mu_C}{2k_BT}\right] + \frac{e^2}{4\hbar}\left[\frac{1}{2}+\frac{1}{\pi}\arctan\left(\frac{\hbar\omega-2\mu_C}{2k_BT}\right) - \frac{j}{2\pi}\ln\frac{(\hbar\omega+2\mu_C)^2}{(\hbar\omega-2\mu_C)^2+(2k_BT)^2}\right] \quad (5)$$

where the first term is attributed to the intra-band transition and the second term is the inter-band transition. The Graphene Fermi level, $\boldsymbol{\mu_C}$, can be changed by applying a gate voltage. The approximate relationship can be estimated as a parallel capacitor. The capacitor-induced carrier concentration of the

IOP Conf. Series: Materials Science and Engineering **840** (2020) 012003 doi:10.1088/1757-899X/840/1/012003

Graphene is $n_g = \varepsilon_0\varepsilon_d V_g/ed$ and the Fermi level is $\mu_C = \hbar\, v_f\, (\pi n_g)^{1/2}$. Here the ε_d is the dielectric constant of the insulating layer and *d* is the thickness of the insulating layer.

The bulk conductivity is related to the surface conductivity by $\boldsymbol{\sigma_V = \frac{\sigma_s}{\Delta}}$, where Δ is the thickness of the Graphene layer, which is much smaller than the excitation wavelength. In this context, Graphene is modeled as a very thin (< 4nm) three-dimensional material. The volumetric anisotropic permittivity, ε, can then be calculated if we assume that the Graphene layer has a finite thickness, Δ.

To turn the surface conductivity of a Graphene layer into a uniaxial anisotropic permittivity, the background relative permittivity of a thick material (Graphite), $\boldsymbol{\epsilon_r}$, is introduced. The uniaxial anisotropic permittivity, in-plane and out of the plane components are then given by [20]

$$\boldsymbol{\epsilon_{\parallel} = \epsilon_r + j\frac{\sigma_s}{\epsilon_0 \omega \triangle}} \quad (6)$$

$$\boldsymbol{\epsilon_{\perp} = \epsilon_r}$$

where $\boldsymbol{\epsilon_r}$ can be set at 2.5, [20] which is based on the dielectric constant of Graphite. The real and imaginary refractive index components can then be obtained from $\boldsymbol{n + ik = \sqrt{\epsilon_{\parallel}}}$.

3. Results

Preliminary simulations are performed for a 3rd-order Sierpinski fractal monopole antenna of nanometric dimensions. The Graphene fractal triangle has lateral dimensions 640 nm x 560 nm and a thickness of 3 nm. The smallest triangle from which the fractal is composed has lateral dimensions 1/16 times the 'fundamental triangle'. The fractal triangle is deposited on top of a 10 nm SiO_2 substrate which rests on top of a 10 nm Si substrate. The fractal antenna is fed by a coaxial cable and placed on a ground plane that has a diameter of 800 nm and a thickness of 25 nm. This is the basic configuration.

Macroscale Sierpinski fractal triangle monopole antennas are intended to be used for different discrete frequency bands. Nanoscale equivalent antennas are not expected to operate in the same way, as the materials have their properties and cannot be assimilated to be a Perfect Electric Conductor (PEC). At the nanoscale, the volumetric conductivity is described in terms of the real and imaginary components of the relative permittivity, which can be translated in terms of the real (n) and imaginary (k) components of the refractive index.

In the model, a Graphene fractal is connected through a coaxial lumped port to a ground plane made of gold. One spherical Perfectly Matched Layer (PML) then includes the Sierpinski fractal triangle, the ground plane and the surrounding volume of air. The bottom of the Sierpinski fractal triangle is flat and connected to the coaxial cable feed. The inner coaxial cable is taken to be gold and surrounded by a SiO_2 outer coaxial casing. In our simulations, the ground plane is also made of gold and the Sierpinski fractal triangle made of Graphene.

As Graphene is two-dimensional material, the volumetric conductivity is expressed in terms of the surface conductivity divided by the Graphene layer thickness as described in Eq. (7). The Sierpinski fractal triangle is positioned on a layer of a dielectric SiO_2 on top of a semiconductor Si substrate. Additional matching parts are not required for using the antenna at higher-order resonances with the given fractal structure. The governing equation for the propagation of electromagnetic waves in the (temporal) frequency domain is obtained from Maxwell's equation and given by

$$\boldsymbol{\nabla x(\nabla x E) - k_0^2 \varepsilon_r E = 0} \qquad (7)$$

$$\boldsymbol{\varepsilon_r = (n - i\,k)^2}$$

For the lumped port, the governing equation is $Z=V_1/I_1$.

The temporal frequency domain model is appropriate under the assumption that all material properties are constant for field strength and that the field will change sinusoidally in time over the range of frequencies considered. The properties of the materials do not play any role in the macroscopic model, providing the conductivity is high enough to qualify as PEC. However, at the nanoscale, the material models for n and k selected in the design change the resonances of the antenna. For the nanometric dimensions, and, given fractal antenna theory the macroscopic Sierpinski, the second and third-order resonances of the small structures are placed at 320 and 600 THz [8]. We study the small structure for 10 -1000 THz.

Concerning the material properties, for the case of Gold, experimental values for n and k are provided for different configurations (films of different thickness, evaporated gold, template stripped gold, single-crystal gold) over the range of wavelength 0.3-25 μm [21], [22]. For Silicon, the n, k data are obtained over the range 0.2-1.45 μm from [23] and [24]. Above this wavelength, the values are extrapolated linearly. For Silicon Dioxide, the SiO_2 values of n and k are available for 0.21 – 1000 μm from [25]. In the case of Graphene, experimental measurements of $\boldsymbol{n}$ and $\boldsymbol{k}$ are available in the region 0.15-0.95 μm from [15]. These values depend on the Graphene thickness and substrate. Alternatively G,raphene surface conductivity is modeled using the RPA in the local limit [12] from Eq. (6), based on validation of the carrier relaxation time measured from the Graphene bulk conductivity at 300 K [10].

Figure 1 presents the model for real and imaginary components of the Graphene relative permittivity, refractive index, and surface conductivity obtained using the RPA in the local limit (Eq. 6) with μ_C= 0.43 eV, corresponding to a carrier concentration of n_g = 1.1× 10^{17} and μ_C=0.19 eV (V_g=2.5V), corresponding to a carrier concentration of n_g =2.2×10^{16} m^{-2} (V_g=0.5V). Graphene has the advantage of tunability, i.e. the opportunity of changing *n, k* by applying a voltage across the thin layer, as this introduces a change on the Fermi energy level ($\boldsymbol{\mu_C}$) and thus of the carrier concentration, n_g. This gives the ability to reconfigure the frequencies of the antenna. Thus, we used values of carrier mobility of $\boldsymbol{\mu}$ = 2000 $cm^2/(Vs)$, $\boldsymbol{v_f}$=1.2 × 10^6 ms^{-1} and $\boldsymbol{\mu_C}$=0.19 eV and 0.43 eV, corresponding to a carrier concentration of n_g =2.2×10^{16} m^{-2} and 1.1× 10^{17} m^{-2}, respectively.

Assuming a SiO_2 of thickness d =10 nm, it takes V_g ~0.5 - 2.5 V to change the Fermi level from 0.19 - 0.43 eV. These values are used to match the experimental validation of a Graphene single layer surface conductivity measurements reported in [10]. The real and imaginary parts of the refractive index of Graphene can be obtained from the surface optical conductivity model as described in Eq. (6) and the volumetric anisotropic permittivity as given in Eq. (7). This model is validated from 3 to 10 μm and can be extended to lower wavelengths. The application of a voltage across the Graphene layer is not modeled, but values of n and k are modified accordingly.

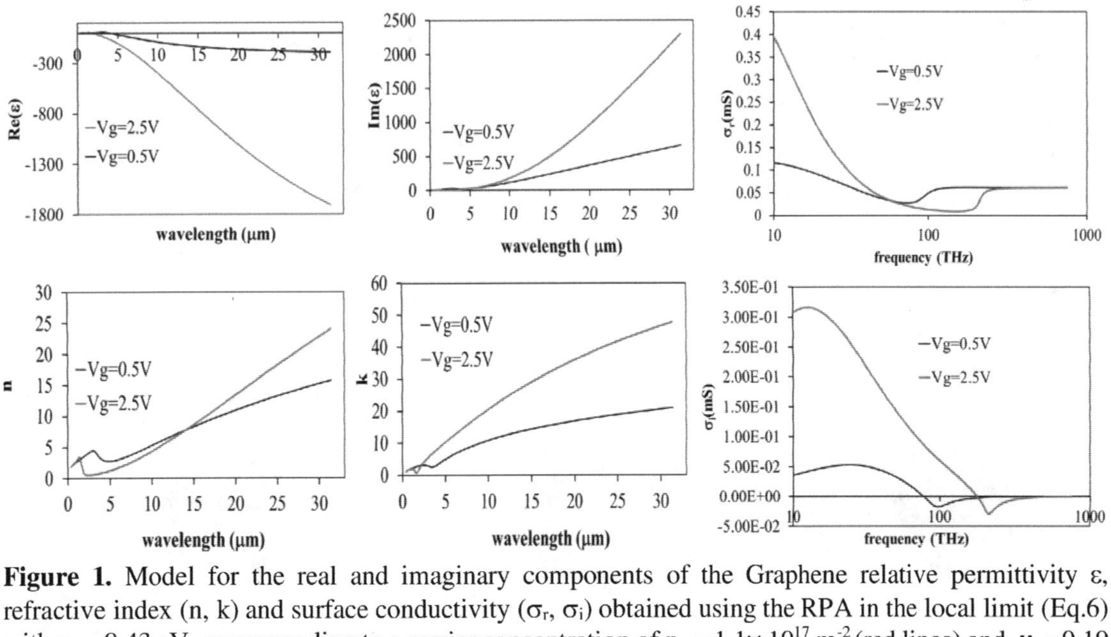

Figure 1. Model for the real and imaginary components of the Graphene relative permittivity ε, refractive index (n, k) and surface conductivity (σ_r, σ_i) obtained using the RPA in the local limit (Eq.6) with μ_C= 0.43 eV, corresponding to a carrier concentration of n_g = 1.1× 10^{17} m^{-2} (red lines) and $\boldsymbol{\mu_C}$=0.19 eV (Vg=2.5V), corresponding to a carrier concentration of n_g =2.2×10^{16} m^{-2} (Vg=0.5V) (blue lines)

Example results of our simulations are given in Figure 2. In this simplified model, the far-field emission is similar to the one that could be achieved by a radiator modeled as a PEC. However, lower resonance frequencies are observed due to the properties of the material. From actual nanometric dimensions but with classic macroscopic Sierpinski fractal antenna theory, second and third-order resonances of the structure are placed at 320 to 600 THz. Depending on the values of n and k for the actual material selected for the fractal radiator the ground plane and inner coaxial, this pattern is changed at the nanoscale as observed in the results given in Figure 2(c). In this case, three resonances are found at 35, 110 and 215 THz.

Current dimensions are dictated by the thickness of Graphene being less than 3 nm. The thickness of Graphene also influences the tunability. This dictates the minimum size of the mesh for modeling the fractal which must be a fraction of the thickness. To keep the number of mesh elements to a minimum, the other dimensions have also been kept to a minimum; hence, the 10 - 1000 THz range, as shown in Figure 2. By further scaling the antenna to operate in the range of frequency 10 - 120 THz,

Graphene may have the advantage of being able to tune the refractive index (surface conductivity) if the thickness is approximated to a single layer or a few layers, by applying a voltage across the layer and substrates. As discussed in [26], the opportunity to model the properties of Graphene given a suitable Maxwell equations solver relies on special conditions to be enforced for the current density within the radiator. This can be accomplished by evaluating the complex permittivity of Graphene as the frequency and other parameters are varied and treating the film thickness as an impedance boundary that establishes a relationship between the currents flowing on the two film surfaces. This increases the number of degrees of freedom and computational complexity, but, in this way, the total losses through the film can be computed with a more computationally stable discretization. Moreover, nanoscale effects such as interface roughness can be introduced, evaluated, and correlated with experimental evidence. Hence, back-to-back modeling and experimental activity are certainly necessary to further progress the design.

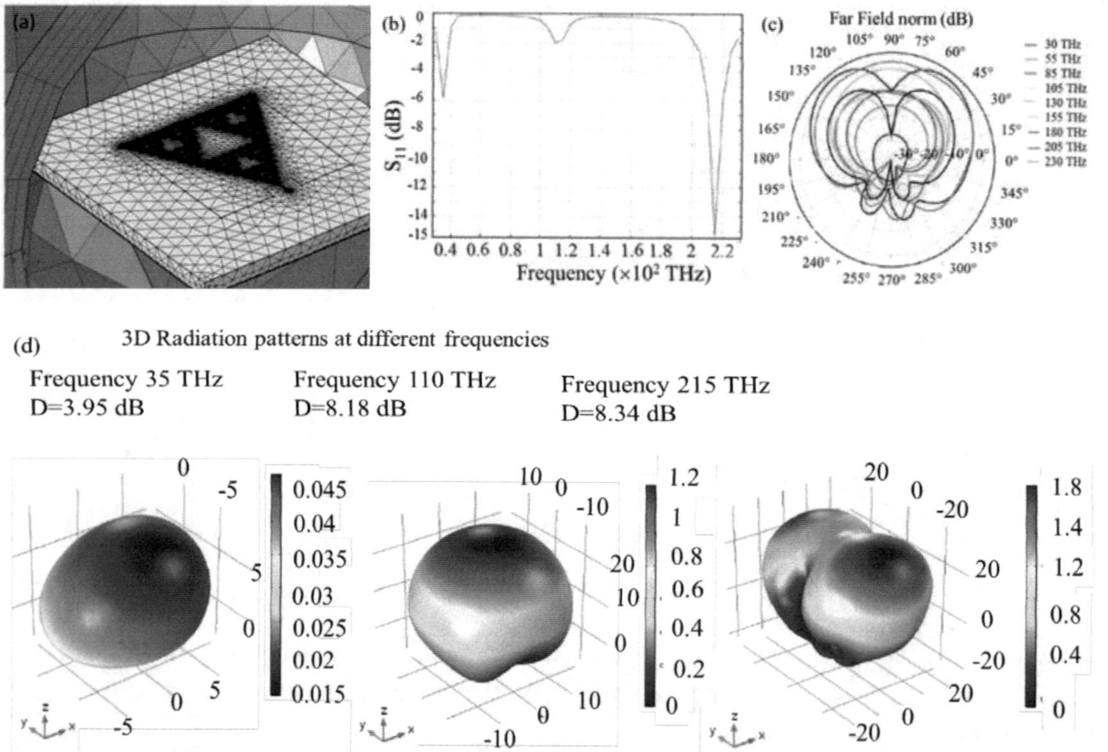

IOP Conf. Series: Materials Science and Engineering **840** (2020) 012003 doi:10.1088/1757-899X/840/1/012003

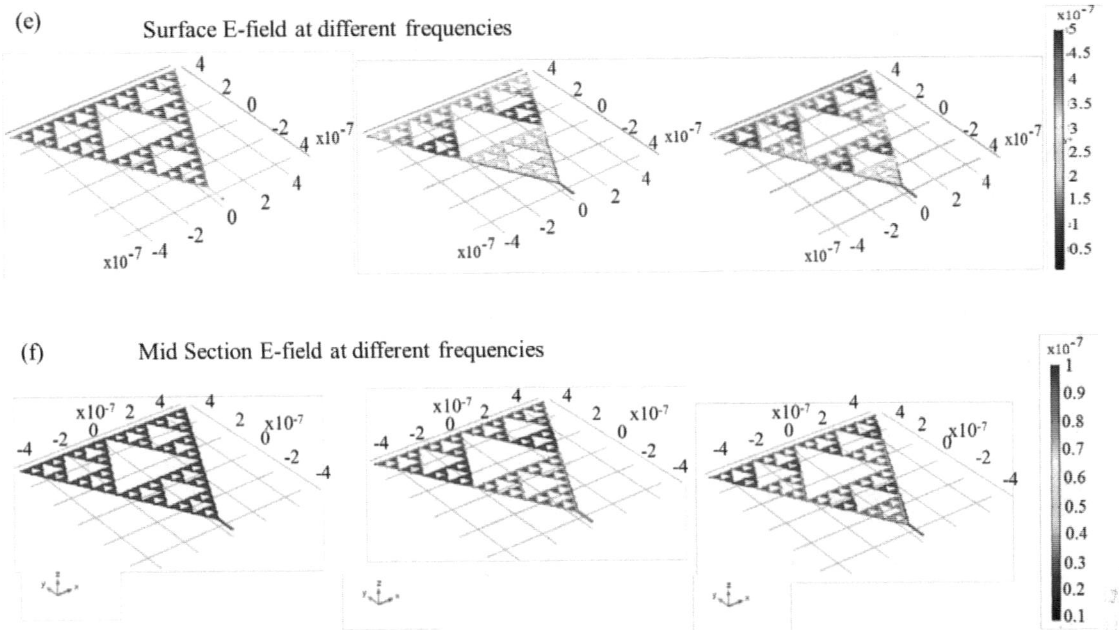

Figure 2. (a) Sierpinsky antenna model. (b) Sierpinski Global S-parameter (in dB) vs. frequency with three resonances at 35, 110 and 215 THz. (c) Sierpinski far-field polar coordinate patterns at selected frequencies. (d) 3D emission patterns. (e) Sierpinski surface E-fields at the resonance frequencies. (f) Sierpinski mid-section E-fields at the resonance frequencies.

4. Conclusions

A preliminary study has been performed for a Graphene fractal antenna to work in the THz frequencies with the opportunity to modulate the emission. The paper has proposed the conceptual background for the study, as well as the result of simulations based on the Maxwell equations (Eq. 8). The model must be carefully revised to simulate the proposed antenna by using externally performed simulations for the conditions to be enforced in the Maxwell solver.

5. References

[1] Geim, A.K.; Novoselov, K.S. The rise of graphene. *Nat Mater* 2007, *6*, 183-191.

[2] Wang, Z.L. Towards self-powered nanosystems: From nanogenerators to nanopiezotronics. *Advanced Functional Materials* 2008, *18*, 3553-3567.

[3] Jornet, J.M.; Akyildiz, I.F. Graphene-based plasmonic nano-antenna for terahertz band communication in nanonetworks. *IEEE Journal on Selected Areas in Communications* 2013, *31*, 685-694.

[4] Ju, L.; Geng, B.; Horng, J.; Girit, C.; Martin, M.; Hao, Z.; Bechtel, H.A.; Liang, X.; Zettl, A.; Shen, Y.R., *et al.* Graphene plasmonics for tunable terahertz metamaterials. *Nat Nano* 2011, *6*, 630-634.

[5] Llatser, I.; Kremers, C.; Cabellos-Aparicio, A.; Jornet, J.M.; Alarcón, E.; Chigrin, D.N. Graphene-based nano-patch antenna for terahertz radiation. *Photonics and Nanostructures - Fundamentals and Applications* 2012, *10*, 353-358.

[6] Zakrajsek, L.; Einarsson, E.; Thawdar, N.; Medley, M.; Jornet, J.M. Lithographically defined plasmonic graphene antennas for terahertz-band communication. *IEEE Antennas and Wireless Propagation Letters* 2016, *15*, 1553-1556.

[7] Dragoman, M.; Muller, A.A.; Dragoman, D.; Coccetti, F.; Plana, R. Terahertz antenna based on graphene. *Journal of Applied Physics* 2010, *107*, 104313.

IOP Conf. Series: Materials Science and Engineering **840** (2020) 012003 doi:10.1088/1757-899X/840/1/012003

[8] Mishra, R.K.; Ghatak, R.; Poddar, D.R. Design formula for sierpinski gasket pre-fractal planar-monopole antennas [antenna designer's notebook]. *IEEE Antennas and Propagation Magazine* 2008, *50*, 104-107.

[9] Xu, Y.Y.; Xu, Y.; Hu, J.; Yin, W.Y. In *Design of a novel reconfigurable sierpinski fractal graphene antenna operating at thz band*, 2013 IEEE Antennas and Propagation Society International Symposium (APSURSI), 7-13 July 2013, 2013; pp 574-575.

[10] Yao, Y.; Kats, M.A.; Genevet, P.; Yu, N.; Song, Y.; Kong, J.; Capasso, F. Broad electrical tuning of graphene-loaded plasmonic antennas. *Nano Letters* 2013, *13*, 1257-1264.

[11] Rosa, L.; Sun, K.; Juodkazis, S. Sierpin´ski fractal plasmonic nanoantennas. *physica status solidi (RRL) – Rapid Research Letters* 2011, *5*, 175-177.

[12] Falkovsky, L.A.; Pershoguba, S.S. Optical far-infrared properties of a graphene monolayer and multilayer. *Physical Review B* 2007, *76*, 153410.

[13] Hanson, G.W. Dyadic green's functions and guided surface waves for a surface conductivity model of graphene. *Journal of Applied Physics* 2008, *103*, 064302.

[14] Firsova, N.E.; Yu, A.F. Quantum description of nanoantenna properties of a graphene membrane. *EPL (Europhysics Letters)* 2016, *114*, 57003.

[15] Ochoa-Martinez, E.; Gabas, M.; Barrutia, L.; Pesquera, A.; Centeno, A.; Palanco, S.; Zurutuza, A.; Algora, C. Determination of a refractive index and an extinction coefficient of standard production of cvd-graphene. *Nanoscale* 2015, *7*, 1491-1500.

[16] Hwang, C.; Siegel, D.A.; Mo, S.-K.; Regan, W.; Ismach, A.; Zhang, Y.; Zettl, A.; Lanzara, A. Fermi velocity engineering in graphene by substrate modification. 2012, *2*, 590.

[17] Han, M.Y.; Özyilmaz, B.; Zhang, Y.; Kim, P. Energy band-gap engineering of graphene nanoribbons. *Physical Review Letters* 2007, *98*, 206805.

[18] Simsek, E. A closed-form approximate expression for the optical conductivity of graphene. *Optics Letters* 2013, *38*, 1437-1439.

[19] Falkovsky, L.A.; Varlamov, A.A. Space-time dispersion of graphene conductivity. *The European Physical Journal B* 2007, *56*, 281-284.

[20] Gao, W.; Shu, J.; Qiu, C.; Xu, Q. Excitation of plasmonic waves in graphene by guided-mode resonances. *ACS Nano* 2012, *6*, 7806-7813.

[21] Gao, L.; Lemarchand, F.; Lequime, M. Comparison of different dispersion models for single layer optical thin film index determination. *Thin Solid Films* 2011, *520*, 501-509.

[22] Olmon, R.L.; Slovick, B.; Johnson, T.W.; Shelton, D.; Oh, S.-H.; Boreman, G.D.; Raschke, M.B. Optical dielectric function of gold. *Physical Review B* 2012, *86*, 235147.

[23] Green, M.A. Self-consistent optical parameters of intrinsic silicon at 300k including temperature coefficients. *Solar Energy Materials and Solar Cells* 2008, *92*, 1305-1310.

[24] Chandler-Horowitz, D.; Amirtharaj, P.M. High-accuracy, midinfrared (450cm−1$\leqslant\omega\leqslant$4000cm−1) refractive index values of silicon. *Journal of Applied Physics* 2005, *97*, 123526.

[25] Kitamura, R.; Pilon, L.; Jonasz, M. Optical constants of silica glass from extreme ultraviolet to far infrared at near room temperature. *Appl. Opt.* 2007, *46*, 8118-8133.

[26] COMSOL Webinar "Simulating Graphene-Based Photonic and Optoelectronic Devices", cn.comsol.com/community/exchange/361/

IOP Conf. Series: Materials Science and Engineering **840** (2020) 012004 doi:10.1088/1757-899X/840/1/012004

Electronic properties of [111] hydrogen passivated Ge nanowires with surface substitutional lithium

L G Arellano[1], F Salazar[1*], A Trejo Baños[1], A Miranda[1], L A Pérez[2] and M Cruz-Irisson[1]

[1]Instituto Politécnico Nacional, ESIME Culhuacán, Av. Santa Ana 1000, 04430 Ciudad de México, México
[2]Instituto de Física, Universidad Nacional Autónoma de México, A.P. 20-364, 01000, México, Ciudad de México, México

*Email: fsalazarp@ipn.mx

Abstract. In this work, a density functional theory study of the lithium (Li) effects on the properties of hydrogenated germanium nanowires (H-GeNWs) is developed. In particular, the electronic band structures, densities of states, formation energies, and Li binding energies of H-GeNWs grown along the [111] crystallographic direction with a diamond structure for different concentrations of surface substitutional Li atoms were studied. Ge nanowires with hexagonal cross sections and three different diameters were considered. The results indicate that all studied H-GeNWs maintain a semiconducting behaviour and the size of the energy band gap is a function of the diameter and the concentration of substitutional surface Li atoms. The formation energy analysis reveals than the energy stability of the nanowires increases when the nanowire diameter and the concentration of Li atoms augment. The results of this work give insight of how the electronic properties of H-GeNWs change during the charging process and open the possibility to incorporate them as electrodes in Li-ion batteries.

1. Introduction

The study and understanding of the properties of nanostructured materials has allowed the development of new technologies and promise many more applications. The semiconducting nanowires have been used to build sensing devices that are successfully employed in the detection of diseases or harmful gases [1], optoelectronic devices, and recently in storage energy devices for clean energy sources [2]. Nowadays it is known that it is possible to control some of their physical properties taking advantage of the quantum confinement effect and their large surface [3]. On the other hand, the Li-ion batteries offer the best performance for portable devices, and the focus of the scientific community is the increment of their efficiency, gravimetric capacity, useful life, power and safety for users, in order to build batteries capable to move a vehicle without fossil fuels, as well as large energy storage-supply systems and small devices that can work for a long time inside the human body. Recent theoretical and experimental investigations reveal that Ge nanowires can be incorporated as anodic materials in the new generation of batteries, [4,5] since bulk Ge has a theoretical capacity of 1200 mAhg-1, a good conductivity due to its small gap, a low working potential respect Li/Li+, and a better Li diffusion in comparison with the bulk Si [6,7]. In this work, a DFT study of the electronic properties of hydrogen passivated Ge nanowires with surface Li atoms as a function of the nanowire diameter and Li concentration is shown.

IOP Conf. Series: Materials Science and Engineering **840** (2020) 012004 doi:10.1088/1757-899X/840/1/012004

2. The model

The nanowires structures were modelled according to the supercell scheme and relaxed using the forces obtained from DFT calculations within the local density approximation with the norm conserving pseudopotentials [8] incorporated in the SIESTA code [9]. A double-ξ s, p basis with a single d orbital was used. All the studied nanowires were free to relax until the Hellmann–Feynman forces were lower than 2 meV/Å. The real space grid for numerical integrations was defined with an energy cutoff of 150 Ry. The Brillouin zone of the nanowire unit cell was sampled with 1 × 1 × 4 k-points for the relaxation process and 1 × 1 × 24 k-points for the electronic band and density of states calculations, within the Monkhorst–Pack scheme. To verify the suitability of the pseudopotentials, the bond length (2.44 Å) and the lattice constant (5.64 Å) of bulk Ge diamond structure were calculated with an error lower than 0.8% for both values [10]. Using periodic boundary conditions for the unit cell geometries and considering a lateral separation of five nanowire diameters between neighbouring nanowires, an isolated, periodic and infinitely long wire along the [111] crystallographic direction, taken as the z direction, was simulated. Figure 1 shows the top and side views of the non-relaxed morphologies of the H-GeNWs unit cells studied in this work, whose diameters are 7.4 Å, 12.0 Å, and 16.6 Å, which have been labelled as W1, W2, and W3, respectively. The diameter of the nanowires was obtained following the same procedure as in reference [10]. The green, white and violet spheres represent Ge, H and Li atoms, respectively. The numbers indicate the sequence by which the Li atoms substitute the H ones along the nanowire surface. In general, the Li atoms were placed in opposite sides to equilibrate the internal stress. The procedure to replace H atoms by Li ones is as follows: The first Li atom, labelled as 1 in Figure 1 was placed in the relaxed structure of the pure H passivated nanowire, and then the system was structurally optimized. Then another H atom is replaced by a second Li atom, labelled as 2 in Figure 1, and the structure was also relaxed. Two more Li atoms replaced two H atoms, labelled as 4 in Figure 1, in the last relaxed structure, and then the whole system was again structurally optimized. This process continued until reaching a concentration of 6 (for W1) and 12 (for W2 and W3) Li atoms per unit cell.

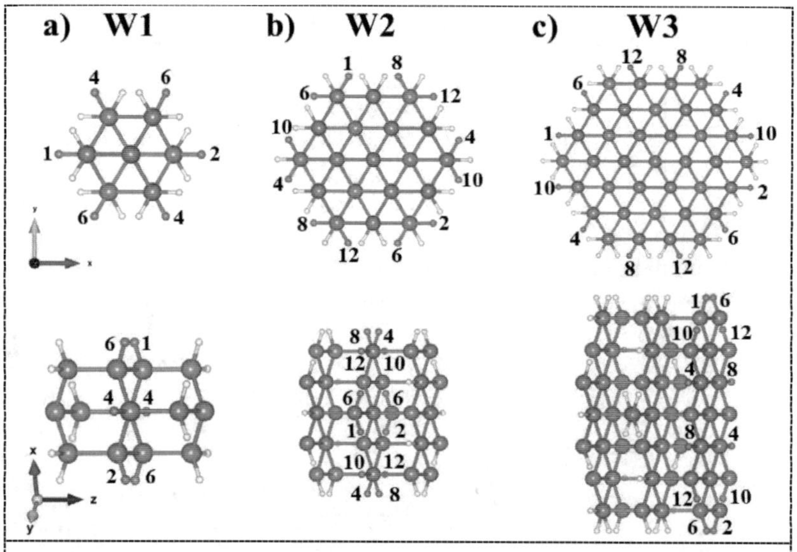

Figure 1. Cross-sectional and side views of the unrelaxed unit cells of [111]-GeNWs considered in this work. Each morphology shown in (a), (b), and (c) is labelled as W1, W2, and W3, respectively. The green, white and violet spheres represent Ge, H, and Li atoms, respectively. The numbers on the Li atoms indicate the sequence by which the Li atoms replaced the H ones.

IOP Conf. Series: Materials Science and Engineering **840** (2020) 012004 doi:10.1088/1757-899X/840/1/012004

3. Results

Figure 2 shows the electronic band structures, the electronic densities of states (DOS) together with the partial DOS of atomic species Ge, H and Li for the relaxed morphologies W1, W2 and W3 of the hydrogen passivated GeNWs. The size of the energy band gap (ΔE_g) diminishes when the diameter of the nanowire augments, as expected from the quantum confinement effect. The type of energy gap is indirect for the W1 and W3 cases, whereas for W2 it is direct.

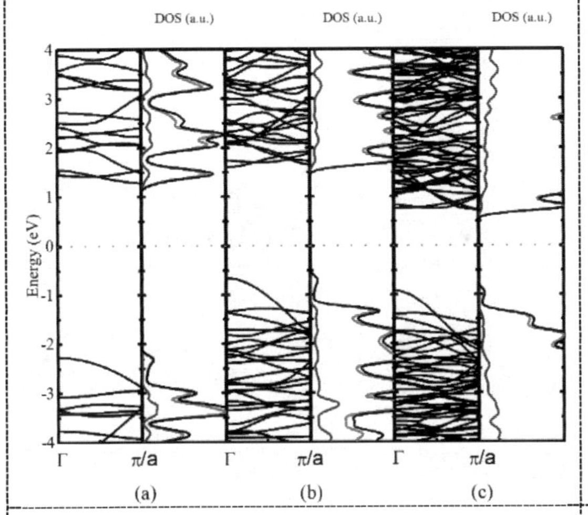

Figure 2. Electronic band structures and densities of states (DOS) of H-GeNWs with diameters (a) 7.4 Å (W1), (b) 12.0 Å (W2) and (c) 16.6 Å (W3). At the right of each band structure, the total (black line) and partial DOS per atomic species Ge (red line) and H (blue line) are shown. The dashed line indicates the Fermi energy (shifted to zero).

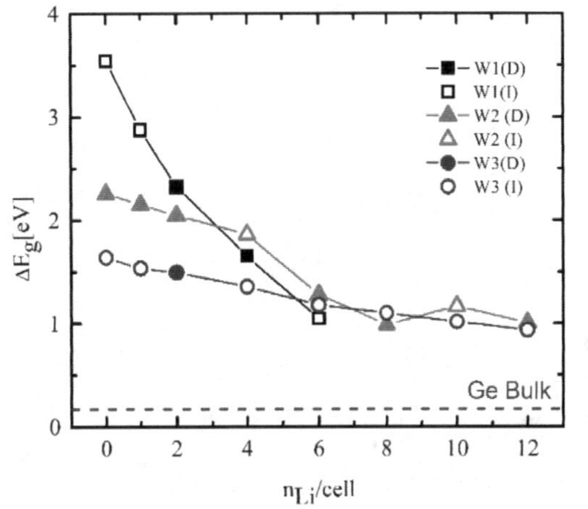

Figure 3. Energy band gap (ΔE_g) of the [111]-GeNWs as a function of the number of Li atoms per unit cell (n_{Li}). The squares, circles, triangles correspond to the W1, W2, and W3 morphologies, respectively. The open (solid) symbols correspond to direct (indirect) ΔE_g. The dashed line corresponds to the gap of bulk Ge calculated with the same level of theory.

Figure 3 shows ΔE_g as a function of the number of substitutional Li atoms per unit cell (n_{Li}). The squares, triangles and circles correspond to the W1, W2, and W3 morphologies shown in Figure 1, respectively. The open (solid) symbols indicate that the band gap is indirect (direct). For all studied morphologies, the band gap decreases as a function of n_{Li}, maintaining their semiconductor behaviour in all cases. Even more, the effect of the surface Li atoms is stronger than that of quantum confinement. The dashed line in Figure 3 indicates the band gap of bulk Ge calculated with the SIESTA code. These results show that the size and nature of the band gap depends on the diameter and the concentration of the Li atoms on the surface of the GeNWs.

The formation energy per Li atom $E_f(n_{Li})$ is an energetic stability parameter of lithiated H-GeNWs and it is given by

$$E_f(n_{Li}) = \frac{1}{M}\left[E(GeNW + [n_H - n_{Li}] + n_{Li}) - n_{Ge}E(Ge_{bulk}) - (n_H - n_{Li})\frac{1}{2}E(H_2) - n_{Li}E(Li)\right]$$

where M is the number of atoms in the unit cell, $E(GeNW + [n_H - n_{Li}] + n_{Li})$ is the energy of the a GeNW unit cell with n_{Li} surface Li atoms and $n_H - n_{Li}$ H atoms, $E(Ge_{bulk})$ is the energy per atom of bulk Ge with diamond structure, $E(H_2)$is the energy of the hydrogen dimer, and $E(Li)$ is the energy of an isolated Li atom. A larger negative value indicates a more stable system. Figure 4 shows $E_f(n_{Li})$

of the lithiated H-GeNWs as a function of n_{Li} for the three studied morphologies. The formation energies decrease as a function of the concentration of n_{Li}, indicating that the nanowires gain stability. Also, the stability of the nanowires increase when the diameter grows.

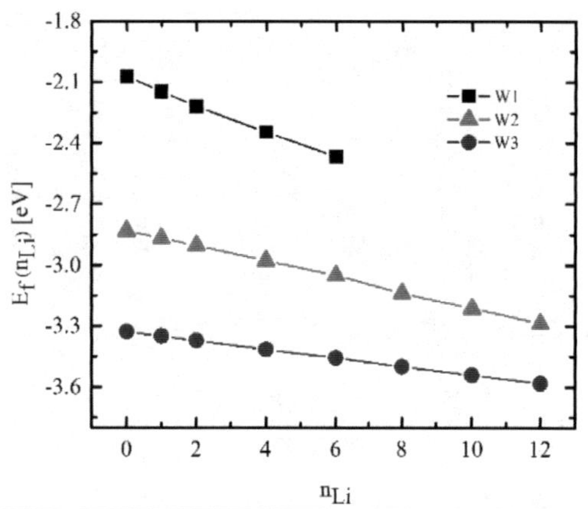

Figure 4. Formation energy (E_f) as a function of the number of Li atoms per unit cell (n_{Li}), for the three morphologies considered in this work.

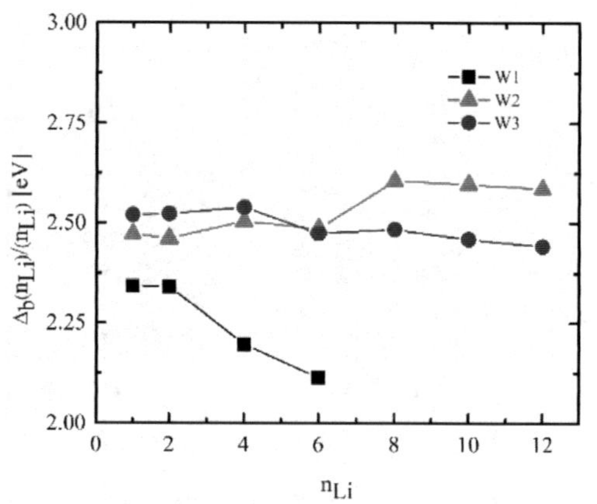

Figure 5. Lithium binding energy (Δ_b) per lithium atom as a function of the number of Li atoms per unit cell (n_{Li}), for the three morphologies considered in this work.

The binding energy of lithium ($\Delta_b(n_{Li})$) quantifies how much energy is necessary to extract n_{Li} atoms from the nanowire, and its dependence on the surface coverage is relevant for Li storage applications. For a hydrogenated Ge nanowire with n_{Li} substitutional Li surface atoms per unit cell, $\Delta_b(n_{Li})$ can be calculated as follows:

$$\Delta_b(n_{Li}) = E(H - GeNW) + \left[n_{Li}E(Li) - (n_H - n_{Li})\frac{1}{2}E(H_2)\right] - E(GeNW + [n_H - n_{Li}] + n_{Li})$$

where $E(H - GeNW)$ is the total energy of the former hydrogenated Ge nanowire. With this definition, a larger value of $\Delta_b(n_{Li})$, indicates a stronger binding of Li to the nanowire. Figure 5 shows the Li binding energy per lithium atom ($\Delta_b(n_{Li})/n_{Li}$) as a function of the number of lithium atoms per unit cell, for the three morphologies considered in this work. Notice that for the nanowires with morphologies W2 and W3, $\Delta_b(n_{Li})/n_{Li}$ is close to 2.5 eV and practically independent of the Li content. These values are slightly higher than those reported in silicene with Li [11] and similar to those reported in previous theoretical works [12]. The thinner nanowire (W1) presents lower $\Delta_b(n_{Li})$ values, indicating a less strong binding with the Ge atoms. The relaxed morphologies reveal a deformation in the structure but all the Ge-Ge bonds still remain. This result contrasts with the case of interstitial Li atoms in nanowires, where the structure usually suffers a strong deformation [12].

4. Summary

In this work, a theoretical study of the effects of substitutional surface Li atoms on the electronic properties of [111]-GeNWs was presented. The results indicate that all studied nanowires maintain a semiconductor behaviour. The band gap size and the energetic stability depend on both the nanowire diameter and the Li concentration. The latter has the strongest effect on the former quantities. Finally, the Li binding energy per atom is almost independent of the concentration of Li atoms having slightly lower values for the smallest studied nanowire. These results reveal that the diameter of the nanowire has little effect on the Li binding energy per atom. This work helps to understand how the nanowire electronic properties are affected upon the insertion of surface Li atoms during the charge process of a battery. It is worth mentioning that the effects of interstitial Li atoms on the electronic properties of GeNWs will be studied in the future. Also the effects of both surface and interstitial Li on the mechanical behaviour of GeNWs should be addressed in order to have a more complete picture of the charging process in these potential anodic materials.

5. References

[1] De Santiago F, Miranda A, Trejo A, Salazar F, Carvajal E, Cruz-Irisson M and Luis A. Pérez L A 2018 *Int J Quantum Chem.* 118 e25713.
[2] Yu K, Pan X, Zhang G, Liao X, Zhou X, Yan M, Xu L, Mai L 2018 *Advanced Energy Materials* 8 32-1802369.
[3] Salazar F and Pérez L A 2012 Comput. Mater. Sci. 63 47–51
[4] Hao J, Wang Y, Guo Q, Zhao J, and Li Y 2019 *Particle & Particle Systems Characterization* 36 9-1900248.
[5] Kennedy T, Mullane E, Geaney H, Osiak M, O'Dwyer C and Ryan K M 2014 *Nano Lett.* 14 716–723
[6] De Santiago F, González J E, Miranda A, Trejo A, Salazar F, Pérez L A, and Cruz-Irisson M 2019 *Energy Storage Materials* 20 438-445.
[7] Salazar F, Trejo-Baños A, Miranda A, Pérez L A, and Cruz-Irisson M 2019 *Journal of molecular modeling* 25 11-338.
[8] Troullier N and Martins J L 1991 *Phys. Rev.* B 43 1993–2006
[9] Soler J M, Artacho E, Gale J D, García A, Junquera J, Ordejón P and Sánchez-Portal D 2002 J. *Phys.: Condens. Matter.* 14 2745–78
[10] Kittel C 1996 *Introduction to solid state physics* John Wiley & sons New York p 23
[11] Osborn T H and Farajian A A 2012 J. *Phys. Chem.* C 116 22916–20
[12] González M A, Salazar F, Miranda A, Trejo B A, Pérez L A, Carvajal, E and Cruz-Irisson M 2018 *Nanotechnology* 29 15- 154004.

Acknowledgments
This work was supported by projects from SIP-IPN, and UNAM-PAPIIT IN109320. Computations were performed at the supercomputer Miztli of DGTIC-UNAM (Project LANCAD-UNAM-DGTIC-180). The authors also thank computing resources, technical advice and support provided by Laboratorio Nacional de Supercómputo del Sureste de México (LNS), through project No. 201903082N, and ABACUS of CINVESTAV. L.G.A. would like to thank BEIFI-IPN, CONACYT and COFAA-IPN for their financial support.

IOP Conf. Series: Materials Science and Engineering **840** (2020) 012005 doi:10.1088/1757-899X/840/1/012005

Decomposition of the Vector Field of Preferred Directions for Optimization of Five-Axis Machining

S. S. Makhanov

Sirindhorn International Institute of Technology, Thammasat University, Rangsit Pathum Thani 12000, Thailand

Email: makhanov@siit.tu.ac.th

Abstract. A new algorithm to increase the production rate of a five-axis milling machine through improving the coordinates of the to-be-milled points transformed from the workpiece to the machine coordinates is presented and analysed. The method optimizes the cutting path by following a vector field (VF) of optimal directions maximizing the material removal rate (MRR). The algorithm includes grid generation, space filling curves (SFC) and a VF decomposition using rotation invariant complex moments. The case of a radial tool path requires a special treatment called Compact Radial Zigzag (CRZ). To reduce the redundancy, the CRZ is composed of layers with a varying step between the tracks. The combination of the proposed techniques generates tool paths which produce complex shaped Stereolithography (STL) surfaces faster than the conventional methods.

1. Introduction

A five-axis machine is a programmable mechanism fed with the CNC program (G-code) i.e. a sequence of commands composed of three Cartesian coordinates of the tool-tip in the machine coordinate system and two rotation angles required to establish the orientation of the tool (Fig 1).

Matching a five-axis toolpath with a VF of preferred directions (VFPD) has been analysed in many publications and is becoming popular in the five-axis machining industry. However, the majority of the algorithms obtain toolpaths by propagating an initial curve inside the parametric region. Optionally, the region is subdivided into clusters characterized by homogeneous VFs. Most of the algorithms have not been verified on industrial formats such as STY, IGES, or STEP.

The VFPD is machine-dependent and is evaluated using the kinematics transformations based on coordinate systems shown in figure 1. The idea to match the tool path and the VFPD based on the maximization of the machining strip was proposed by Chiou and Lee [1]. The first algorithm to cluster the VFPD was introduced in [2]. Some recent modifications are the tensor based approach [3] and segmentation into regions of similar pathlines [4]. A comprehensive survey of these techniques is given in [5] (see also [6],[7]).

When a VFPD-method offsets an initial path and propagates it inside the region, the path may substantially deviate from the VFPD. Then the algorithm re-initializes it.

This procedure often creates configurations which does not actually match the desired VFPD. The tool paths often become redundant i.e. the distance between the tracks is considerably smaller than the maximum allowed machining strip. Besides, these algorithms are sensitive to the choice of the initial path which itself is a computationally expensive, NP-hard problem. Finally, as mentioned above, only a few algorithms have been verified on real industrial formats such as STL, IGES or STEP [8].

In this paper a new VFPD-type toolpath generation method based on curvilinear grids [9]-[11] and Adaptive Space Filling Curves (ASFC) [12]-[13] has been proposed. The method has been applied to the most popular industrial format-STL. The resulting VF-aligned path combines the adaptivity of path-propagating methods with a convenient, regular topology of the conventional patterns such as the zigzag or the spiral. The VF can be obtained using a variety of utility functions such as the machining time, toolpath length (machine independent), kinematic error, etc. In this paper the algorithm constructs a curvilinear path which partly or even entirely aligns with the VF based on the maximum material removal rate MMRR. The reference methods are the standard iso-parametric path (ISO), MasterCam and advanced toolpath generation algorithms of NX11 (former UG) e.g. Helical/Spiral (HS) and Follow Periphery (FP). The cutting experiments have been performed on a milling machine HAAS VF2TR.

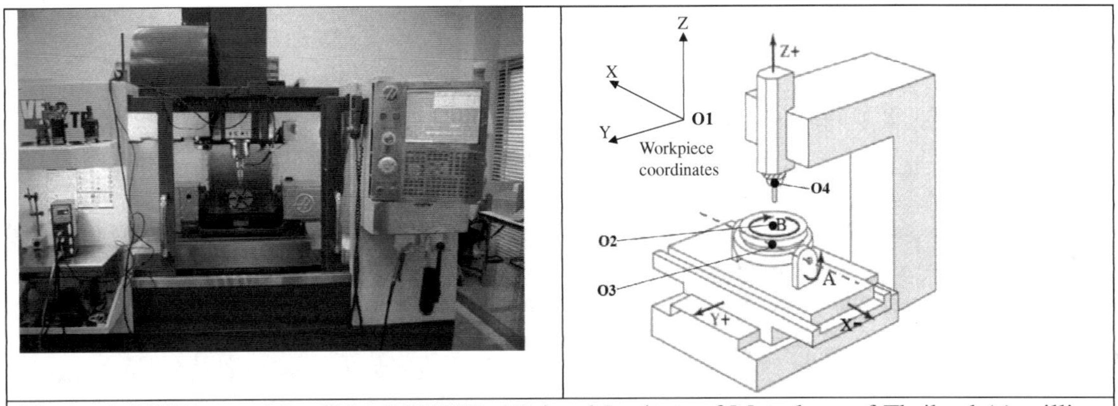

Figure 1. Milling machine HAAS VF2TR, National Institute of Metrology of Thailand (a) milling machine with a high speed spindle, (b) the rotational and translational axis of the machine: O1-workpiece coordinate system, O2-coordinate system of the first rotary part (B-axis), O3-coordinate system of the second rotary part (A-axis), O4-coordinate system of the spindle.

The proposed toolpath generation technology consists of the following steps:

- Flatten the surface (parametrization).
- Select a cost function and constraints to evaluate the optimal directions.
- Evaluate the VFPD.
- Cluster the VFPD using complex moment invariants, and a region growing algorithm.
- For each cluster, generate a toolpath aligned with the VF.
- Extend the boundaries of the clusters.
- Define the order of the machining using the shortest path strategy.

2. Vector field of optimal tool directions

Consider a part surface $S(u,v)$. Let W_1 be a CC (cutter contact) point. Define a collection of points given by $\Omega_{W_1} = \{W : \mathrm{dist}_S(W_1, W) = l_1\}$, where dist_S is the geodesic distance and l_1 is a prescribed step. The image of Ω_{W_1} in the machine coordinates is denoted by Ω_{M_1}. The distance between M_1 and M is denoted by $l_{1,M} \equiv l_{1,M}(W)$.

The machining strip corresponding to $\overrightarrow{W_1,W}$ is denoted by $w_1 \equiv w_1(W)$. Ω_{W_1} and Ω_{M_1} are obtained by the corresponding inverse kinematic transformations [11]. Note that Ω_{W_1} is a geodesic circle, whereas Ω_{M_1} is represented by an irregular closed curve. The distance on $S(u,v)$ is not proportional to the distance in the machine coordinates. The machining time t depends on the translations and rotations of the machine parts necessary to establish the required position and orientation of the tool.

Define the MRR in the direction $\overrightarrow{W_1,W}$ by $R_M(W) = \dfrac{F\, l_1(W) w_1(W)}{l_{1,M}}$ [mm^2/s], where F is the feed rate.

The machining strip w_1 depends on, the curvature of the part surface, the shape of the tool and the effective cutting profile. Methods of evaluating the machining strip are presented in [11]. The direction $\vec{W_1, W_2}$ becomes optimal, if $W_2 = \arg\max_{W \in \Omega_{W_1}} R_M(W)$ e.g. W_2 produces MMRR. Vectors $\vec{W_1, W_2}$ evaluated for each surface point and transferred into the parametric coordinates form a VFPD. Denote VFPD by $V(u,v)$. Unfortunately, a continuous, smooth toolpath maximizing the utility function $R_M(W)$ for every W on a complex shaped part surface usually is not possible. On the one hand, the tool path has to follow the prescribed VF. On the other hand, the desirable geometric structure of such a path is the zigzag, spiral or the radial pattern. The latter requirement improves the smoothness of the machined part, reduces the possibility (although does not guarantee the avoidance) of gouging and global collisions.

3. Grid generation

We arrange the cutter location (CC) points $\{ (u_{i,j}, v_{i,j}),\ 0 \le i \le N_\xi\ \ 0 \le j \le N_\eta \}$ as a curvilinear grid in the parametric coordinates u,v. Mathematically, the CC points define a mapping from the computational region $\Delta = \{0 \le \xi \le N_\xi,\ 0 \le \eta \le N_\eta\}$ into a domain K defined in the parametric coordinates. In other words, there exists a pair of functions $\{u(\xi,\eta), v(\xi,\eta)\}$ such that the rectangular grid $\{i, j\}$ being fed to $\{u(\xi,\eta), v(\xi,\eta)\}$ becomes $\{u_{i,j}, v_{i,j}\}$

According to this model, the required VF is partitioned into $(\alpha(u,v), \beta(u,v))$, creating a dual VF, corresponding to the ξ and η directions (figure 2):

$$\alpha(u,v) = \begin{cases} V(u,v) \in \Omega_\xi, \\ 0, \text{otherwise}, \end{cases} \quad \beta(u,v) = \begin{cases} V(u,v) \in \Omega_\eta, \\ 0, \text{otherwise}, \end{cases}$$

where Ω_ξ, Ω_η are prescribed subsets of $V(u,v)$ selected according to a certain criteria.

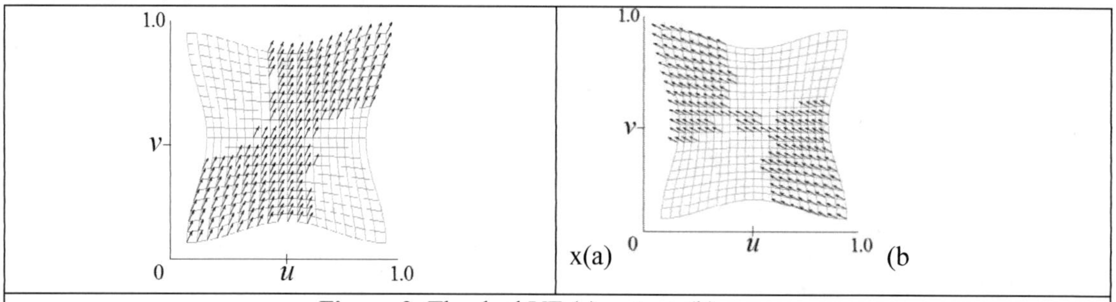

Figure 2. The dual VF (a)- $\alpha(u,v)$,(b)- $\beta(u,v)$

The grid $\{u(\xi,\eta), v(\xi,\eta)\}$ is adapted to the dual VF $(\alpha(u,v), \beta(u,v))$ using a modification of a standard grid generation approach [6]-[9]. The smoothness of the grid is represented by a functional given by

$$F_S = \int\int u_\xi^2 + u_\eta^2 + v_\xi^2 + v_\eta^2 d\xi d\eta,$$

where subscripts denote the partial derivatives. The numerical solution is based on approximating of the second-order elliptic PDEs and an iterative solution by the Newton-Raphson method.

4. Numerical and cutting experiments

This section presents a VF-grid alignment method combined with the ASFC [10], tested against ISO, "Follow Periphery" (FP) and "Helical or Spiral" (HS) routines of NX11. A five-axis toolpath cannot be generated by NX11 directly from the STL format. Therefore, we convert the STL to a solid model, using fit surface functions of NX11. The average Euclidean error is 0.05 mm, which ensures the accuracy of the approximated surface. The HS and FP toolpaths are generated from the solid model. An example of the test surface is shown in figure 3. The blank workpiece is 200x200x50 mm. The curvilinear grid adapted to the dual VF in figure 3 is shown in figure 3(a) and the corresponding tool path in figure 3 (b).

The surfaces produced by the conventional ISO zigzag and by the proposed method are displayed in figure 4 (a) and (b) respectively. Table 1 shows the machining time. The proposed method outperforms the conventional algorithms for every prescribed scallop height, h. For instance, for h=0.25mm the VF tool path is 5.3 times faster than that generated by MasterCam, by 23% faster than ISO, 7.8 times faster than HS and by 40% faster than FP.

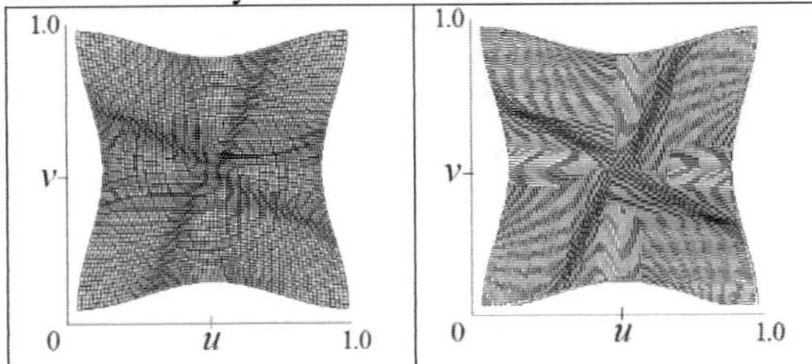

(a) (b)

Figure 3. (a) adaptive grid, (b) SFC- tool path

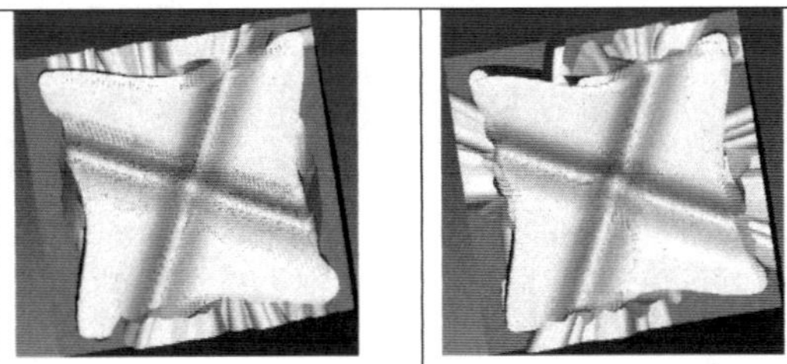
(a) (b)

Figure 4. The machining results (a) ISO zigzag (b) the VF method

5. Decomposition of the VF using flusser complex moments

Several algorithms have been proposed to decompose the VF of optimal directions for tool path generation of five-axis machines e.g. [3],[4]. However, most of the algorithms are not applicable to the STL surfaces since the corresponding VF is usually highly irregular and noisy. In this section we present a clustering approach based on complex moments proposed by Flusser [14] to circumvent the above drawbacks. We illustrate the algorithm by radial (star) pattern which requires a special treatment due to the high redundancy near the pole. The basic idea of the CRZ is minimization of the machining time by decomposing the region into several subregions characterized by an increasing stepover between the radial tracks[15]. A video demo of the algorithm is available at:
https://drive.google.com/open?id=1OM_z4cAOUqGu2RPAzkZOIBcEnfptdTq7 .

The model of the crown of the canine tooth produced by the proposed method is shown in figure 5.

Finally, the machining time for the test examples in figures 3 and 4 is compared with the conventional methods in Tables 1 and 2.

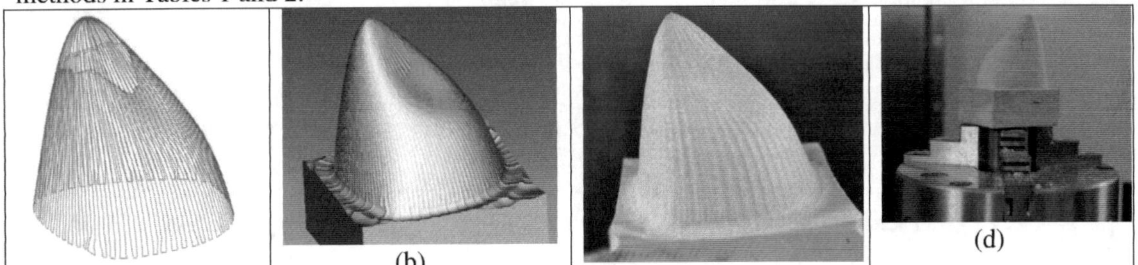
(a) (b) (c) (d)

Figure 5. Application of the CRZ to a dental crown, (a) two-layer CRZ, (b) virtual machining (Vericut), (c) workpiece prepared for the fine cut, (d) the result

6. Conclusions

A new method for generation of VF-aligned tool paths for five-axis machining has been presented and analysed. The new idea is numerical generation of a curvilinear tool path adapted to the VF of optimal directions. This is combined with clustering of the VF and generation of local optimal subgrids providing faster machining. By introducing internal radial layers, the proposed algorithms eliminates the redundant parts. The experiments show a substantial decrease in the machining time with regard to preceding methods. In our experiments the proposed method outperforms the conventional algorithms for every prescribed scallop height, moreover, the advantage increases as the scallop height decreases. The CRZ partition makes it possible to set up the maximum machining strip on every circular boundary, reducing the redundancy, and therefore, reducing the machining time. The process is repeated until the total machining time cannot be improved.

Table 1. Machining time, Surface 1

Method		Scallop height, mm		
		0.25	0.1	0.05
NX11	HS	2:59:36	7:10:23	14:54:49
	FP	1:27:47	2:18:48	3:16:02
ISO zigzag		0:33:49	0:53:10	1:15:04
ASFC		**0:26:43**	**0:41:56**	**0:56:26**

Table 2. Machining time, Surface 2

Method		Scallop height, mm		
		1.0	0.5	0.1
NX11	HS	04:19.	04:30	08:21
	FP	04:20	06:00.	09:01
ISO	Zigzag	01:32	02:09	05:16.
	Contour	03:39	04:00	07:09
	Radial	**01:47**	02:17.	02:32.
Iso-planar		01:58.	02:45.	03:34
CRZ		**01:47**	**01:58**	**02:06.**

7. References

[1] Chiou C and Lee Y-S 2002 *Comp.- Aided Des.* **34** 357-71.
[2] My C A, Makhanov S S, Munlin M, Bohez E J and Tabucanon M 2005 *Int J CAD/CAM* **5** 1–14.
[3] Liu X, Li Y G, Ma S and Lee C-H 2015 *Comp.- Aided Des.***66** 1-13.
[4] Kumazawa G H, Feng H Y and Fard M J B 2015 *Comp.- Aided Des.* **67** 1-12.
[5] Moodleah S, Bohez E J and Makhanov S S 2016 *Int. J. Prod. Res,* **54** 7296-329.
[6] Tang Q, Yin S, Chen F, Huan S, Luo H, and Geng J 2018 *Int.J Adv. Manuf. Tech* **97**: 3523–34.
[7] Hu P, Chen L, and Tang K 2017 *Comp.-Aided Des.*83: 33–50.
[8] Sun Y W, Guo D, Jia M.Z and Wang H X 2006 *Int.J Adv.Manuf. Tech.* **28** 721-6.
[9] Giannakopoulos A E and Engel A J 1988 *J. of Comp. Phys.* **74** 422-39.
[10] Liseikin V D 2010 Grid generation methods Springer.
[11] Castillo J E and Otto J S 2000 *Comp. & Math. with Appl.* **40** 523-35.
[12] Moodleah S and Makhanov S S 2015 *Int. J. Adv. Manuf. Tech.* **80** 65-90.
[13] Makhanov S S and Anotaipaiboon W 2010 Advanced Numerical Methods to Optimize Cutting Operations of Five-Axis Milling Machines, Springer.
[14] Flusser J and Suk T 2006 *IEEE Trans. Image Proc.* **15**, 3784–90.
[15] Dang L V, Vacharanukul K and Makhanov S S 2019, *Int.J. Adv. Manuf. Tech,* **105**, 1853-82.

Acknowledgments
This research is supported by the Center of Excellence in Biomedical Engineering of Thammasat University.

IOP Conf. Series: Materials Science and Engineering **840** (2020) 012006 doi:10.1088/1757-899X/840/1/012006

Accurate Scallop Evaluation Method Considering Kinematics of Five-axis Milling Machine for Ball-end Mill

Luu Hai Tuan[1, a], Stanislav S Makhanov[1, b]

[1]School of Information, Computer and Communication Technology, Sirindhorn International Institute of Technology, Thammasat University, Pathum Thani 12120, Thailand.

Email: [a]xhaituanx@gmail.com, [b]makhanov@siit.tu.ac.th

Abstract. A new algorithm to evaluate the scallops left between consecutive tool tracks after five-axis machining of a complex-shaped part surface has been proposed. The algorithm has been developed for the ball-nose cutter. The novelty of the algorithm includes a variable plane to evaluate the effective tool profile and the part surface profile, the orientation of the tool as well as non-linear kinematics of the five-axis machine. The proposed algorithm has been specifically designed for and tested on the industrial Stereo lithography (STL) format representing complex shaped synthetic five-axis parts and a model of a crown of the molar tooth. The procedure has been tested against several modifications of the conventional curvature based method and the sphere intersection method. The ground truth is generated using the solid modeling engine of Vericut 8.2. The algorithm provides a tangible accuracy increase in terms of the average and the maximum error with regard to the reference methods.

1. Introduction

The scallops between the consecutive tracks of the five axis machining tool is a critical measure of the quality of the part surface. It is a paramount importance that a tool path generation algorithm has been integrated with an accurate scallop evaluation procedure. In many cases, it is imperative for the tool path structure and topology, consequently defines the quality of the output part. The conventional iso-parametric and the iso-planar methods lead to conservative path intervals to control the scallops [1-2]. Only at some selected points, the scallops are compared with the prescribed tolerance. As the result, in some areas the part surface is unnecessarily accurate, and in some other areas the scallops are actually larger than that required by specification. While, having too small scallops is usually due to the redundancy of the tool path, the large scallops usually appear due to the errors in the scallop evaluation procedures integrated into toolpath generation. One of the first scallop evaluation formulas proposed by Suresh and Yang [3] is given by

$$h = r - \left(r^2 - a^2\right)^{1/2} \tag{1}$$

for a flat surface and

$$h = s\left\{-R + \left[(R + sr)^2 - a^2\right]^{1/2}\right\} - \left(r^2 - a^2\right)^{1/2} \tag{2}$$

for a convex $(s = 1)$ and a concave $(s = -1)$ surface, where r is the tool radius, R is the surface curvature, and a is the half distance between two adjacent tool centers. However, the first order Taylor expansion formula 'as in equations (1) and (2)' is suitable only when the converted parametric interval is small. Therefore, Lin and Koren [4] propose a neat evaluation formula given by

IOP Conf. Series: Materials Science and Engineering **840** (2020) 012006 doi:10.1088/1757-899X/840/1/012006

$$h = \frac{w^2(R+sr)}{8rR}, \tag{3}$$

where h is the maximum allowable scallop height, w is CC path interval and $s = 0, -1, 1$ respectively for flat, concave and convex surface. Their technique is based on the second order Taylor series and an error compensation. The above formula has been later used by many authors (see for instance [1, 5-9]). The above methods require evaluation of the curvature between the cutter contact (CC) points and a local approximation of the part surface by a sphere. Another version of this idea is a polynomial surface such as the Bezier patch, NURBS, etc.

A prominent example of the importance of the fast and accurate scallop evaluation is the family of the iso-scallop tool path generation methods (see a comprehensive survey of iso-scallop methods in [10]). However, the majority of the methods do not consider the kinematics of the machine producing non-linear trajectory between each pair of CC points. It is important that the scallop between two consecutive tool positions also changes non-linearly and depends on the particular kinematics of the machine.

Further, the above algorithms are applied at the intersection of the effective tool profile and the part surface in a certain reference plane. Intuitively, the reference plane must contain the vector connecting neighboring cutter contact (CC) points and the vectors corresponding to the tool orientations at these CC points. However, constructing a plane passing through three linearly independent vectors is not possible. Therefore, the plane is defined using certain practical assumptions. For instance, when the surface is an explicit function of the horizontal coordinates, the CNC programmers often consider planes parallel to the z-axis. However, even in this simplest case, the ball-nose cutter produces the same scallop height for totally different orientations of the tool, whereas the actual scallop could be much smaller or larger.

Another version of this model constructs the reference plane passing through the CC point and perpendicular to the feed direction [4]. In this case, the plane is not perpendicular to the feed direction at the second CC point and may not even contain the second CC point. Hence, the algorithm considers a midpoint and an "average" feed direction at the midpoint.

A more sophisticated version of the scallop evaluation considers a set or reference planes rotated at the prescribed point such as the CC-midpoint, applying 'equation (3)' at each plane and considering maximum resulting scallop. However, it is not clear how to accurately evaluate the tool profiles at these plane. At some extreme positions of the planes, the actual 3D bodies of the tool on the consecutive tracks may not even meet. Finally, [11] proposes to evaluate the scallop height for the ball-nose cutter using intersection of corresponding spheres follows a vector formed by center of intersection circle and the middle point between two neighboring CCs. However, the method can be inaccurate in case of sharp curvature variations. Moreover, it has been proposed and verified on a three-axis machine and only on parametric surfaces.

2. Methodology

Our methodology is based on the direct estimation of the intersection of the spheres representing the ball-nose cutter which follows the non-linear trajectories produced by the inverse kinematic transformation of a particular five-axis machine.

2.1. Ball nose tool intersections

Consider a ball nose tool with the diameter d. Consider an arbitrary tool path and two neighboring CL points p_1 and p_2. If $D \equiv \|p_1 - p_2\| > d$ the corresponding spheres do not intersect. Otherwise the intersection is a circle C_{12} with the radius

$$r' = \frac{1}{2}\left(d^2 - D^2\right)^{1/2}. \tag{4}$$

Consider a parametric circle in the yz–plane given by $C_1(t) = \{0, r'\sin t, r'\cos t\}$. Define $v_3 = p_2 - p_1$ A plane P with the normal vector v_3 through an origin is given by $(v_3, M) = 0$, with $M = (x, y, z)$.

Consider two arbitrary linearly independent vectors $v_1, v_2 \in P$. Orthogonalizing v_1, v_2 and normalizing v_3 produces a new orthonormal basis $\{\hat{v}_3, \hat{v}_1, \hat{v}_2\}$ (in this order). Columns of the corresponding transformation matrix T are composed of the above basis vectors. Clearly, in the workpiece coordinates

$$C_{12}(t) = TC_1(t) + \frac{D}{2}\hat{v}_3. \quad (5)$$

The coordinate transformation is illustrated in 'Figure 1(a)', where P is a plane through p_1 and P' is a plane containing $C_{12}(t)$.

2.2. Scallop estimation

The cutter contact (CC) points corresponding to the CL points p_1 and p_2 are given by

$$p'_k = p_k - \frac{d}{2}t_k,\ k = 1,2, \quad (6)$$

where t_1 and t_2 are the tool orientation vectors.

Define the reference plane P_{12} by

$$\left(n_C, M - p'_m\right) = 0, \quad (7)$$

where p'_m is the middle point between p'_1 and p'_2, $n_C = n_m \times v_{CL}$, $v_{CL} = \overrightarrow{p_1 p_2}$ and n_m is a vector approximating the normal to the surface at p'_m. For instance, $n_m = (n_1 + n_2)/2$, where n_1, n_2 are the normal vectors at p_1' and p_2' 'Figure 1b'.

In order to find the intersection points A and B between C_{12} and P_{12}, we use a built-in function "solve" of MATLAB. The function finds two roots s_1 and s_2 so that $A = C_{12}(s_1)$, $B = C_{12}(s_2)$.

Since $\overrightarrow{AB} \in C_{12}$ and $\overrightarrow{AB} \in P_{12}$, the scallop h is defined by the distance between the point B and the part surface along the vector $\overrightarrow{AB}$ 'see Figure 1(b)'. Furthermore, we consider not only fixed CL points p_1 and p_2 but all spheres centered at the trajectories $p_1(t), p_2(s)$ as long as $\|p_1(t) - p_2(s)\| \le d$. 'Figure 1(c)'. In order to evaluate $p_1(t)$, $p_2(s)$, the inverse kinematic transformation of a five-axis machine is employed. The experimental machine is the vertical five-axis machining center Haas VF-2TR [7]. The machine configuration and the scheme of the inverse kinematics are illustrated in Figure 2.

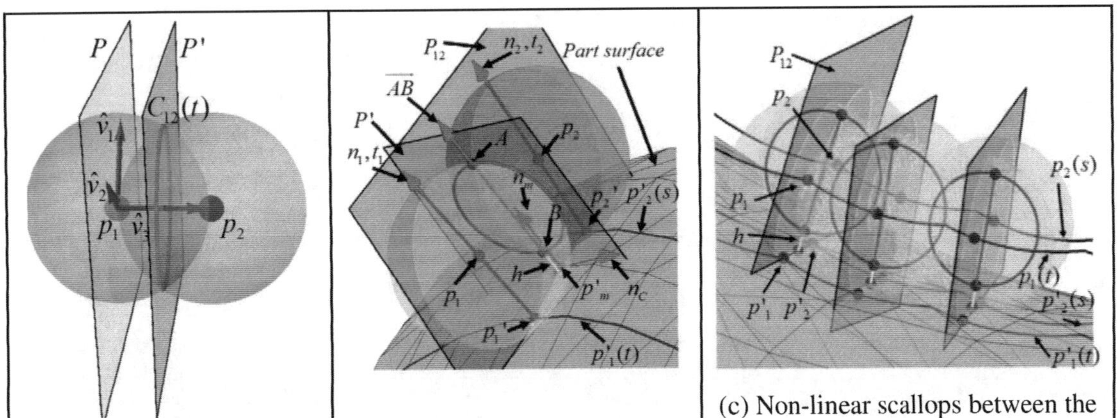

(a) Basis vectors (b) Intersecting spheres (c) Non-linear scallops between the tool trajectories

Figure 1. Ball nose tool intersections and the scallop estimation.

IOP Conf. Series: Materials Science and Engineering **840** (2020) 012006 doi:10.1088/1757-899X/840/1/012006

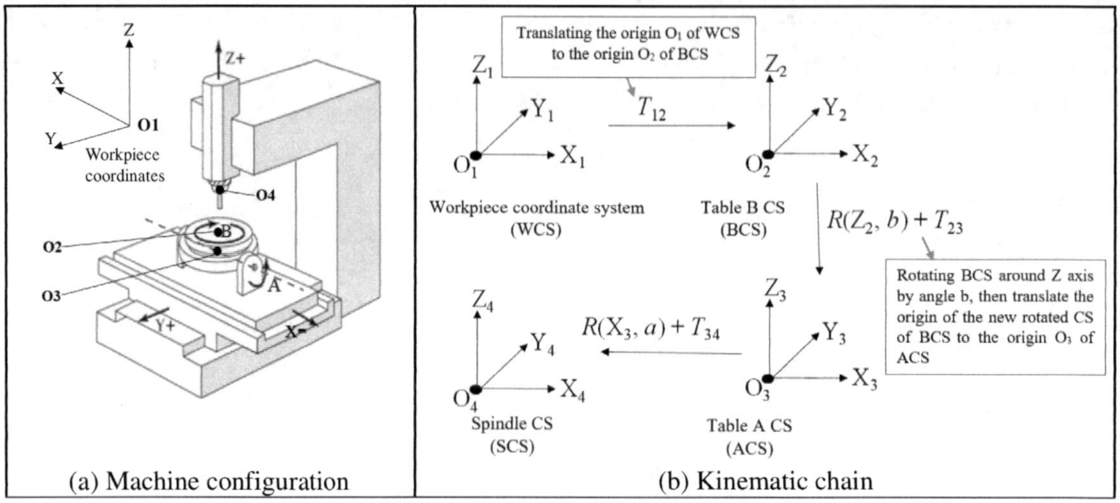

Figure 2. Five-axis milling machine with the rotary axes on the table.

3. Numerical experiments and testing against Vericut-"ground truth"

The proposed method called Spheres/Kinematics (SK) has been tested using a synthetic parametric surface, a synthetic STL surface and an STL model of the molar tooth crown. The reference methods are the curvature based method 'equation (3)' and a method based on intersection of the spheres [11] (see the introduction). The following procedure is employed: first a conventional tool path is generated. Next, the procedure evaluates the scallops between the tool tracks using the proposed method and the two reference methods. Finally, a cutting by the virtual five-axis machining center, Haas VF-2TR is performed by Vericut 8.2. The resulting solid model of the part surface S_V is compared with the actual parametric surface $S(u,v)$ or with the STL model. Surface 1 (Saddle) is given explicitly by $z = x^2/20 - y^2/20 - 31.25$ 'Figure 3(a)'. The size of the base of the part is 50x50 mm. The surface 2 (Waves) is an STL surface depicted in 'Figure 4(a)'. The size of the base of the surface is 50x50 mm. Surface 3 is an STL model of a molar tooth crown in 'Figure 5(a)'. The size of the base of the surface is 60x60 mm. A conventional iso-parametric zigzag was used to cut the surface 1 in 'Figure 3 (b)'. Surface 2 and 3 were cut using iso-planar zigzag path shown in 'Figure 4(b)' and 'Figure 5(b)'. All surfaces were cut by the ball-nose tool having the diameter 4 mm. The resulting solid models produced by Vericut are shown in 'Figures 3, 4 and 5 (c)'.

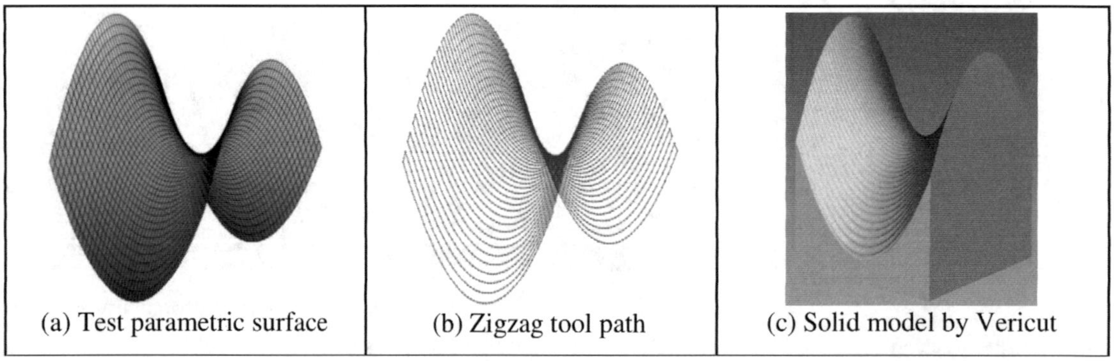

Figure 3. Surface 1 (Saddle).

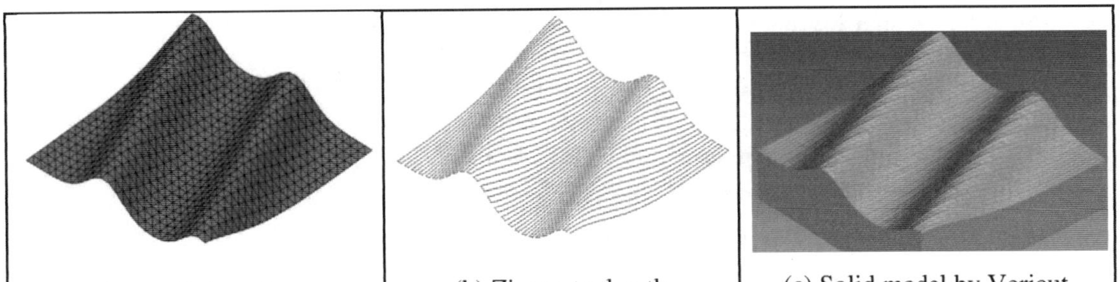

(a) Test STL surface (b) Zigzag tool path (c) Solid model by Vericut

Figure 4. Surface 2 (Waves).

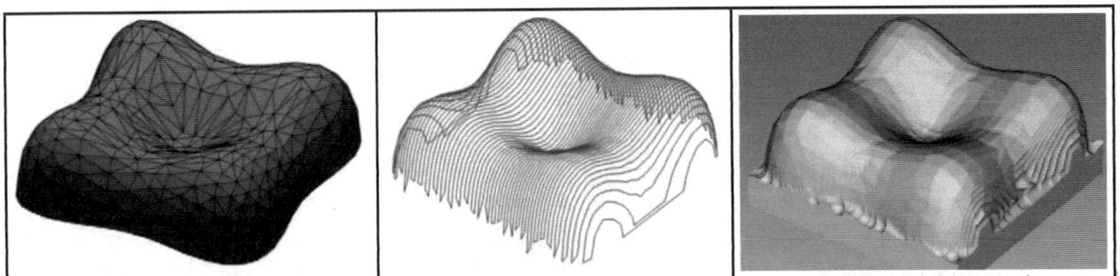

(a) Test STL surface (b) Zigzag tool path (c) Solid model by Vericut

Figure 5. Surface 3 (Molar tooth crown).

Table 1 shows scallop evaluated by the conventional techniques including the curvature method (C-method) and sphere-based method [11] (S-method). Table 2 shows the scallop evaluated by the SK-method and Vericut. The most important are the average and the maximum values.

Next, the scallop evaluation error is defined by

$$\varepsilon_M = \frac{|h_M - h_G|}{h_G} 100\%, \tag{8}$$

where, h_M denotes the scallop obtained by a method M ($M = C, S, SK$ i.e. the curvature method, the spheres method and the proposed sphere-kinematics method respectively) and h_G is the ground truth scallop obtained by measuring the Vericut solid model.

Table 3 shows ε_M produced by SK and the two competing methods. Specifically, SK reduces ε_M in terms of the maximum scallop by 304% with the reference to the C-method and by 39% with regard to the S-method. In turn, SK reduces ε_M in terms of the average scallop by 12% with the reference to the C-method and by 11% with regard to the S-method. The advantage of the SK is explained by a new way of selection of the reference plane, taking into account the non-linear kinematics as well as the tool orientation.

Table 1. Scallop evaluated by conventional methods (mm).

Method/	Curvature method			Spheres method		
Surface	Min	Max	Avr	Min	Max	Avr
Surface1	0.077	0.509	0.218	0.069	0.508	0.216
Surface2	0.030	0.332	0.091	0.024	0.235	0.091
Surface3	0.001	0.592	0.040	0.001	0.188	0.039

Table 2. Scallop evaluated by proposed method and Vericut (mm).

Method/ Surface	SK			Vericut		
	Min	Max	Avr	Min	Max	Avr
Surface1	0.023	0.502	0.123	0.033	0.507	0.161
Surface2	0.020	0.159	0.077	0.020	0.165	0.082
Surface3	0.001	0.176	0.039	0.001	0.137	0.040

Table 3. Scallop evaluation error (%).

Method/ Surface	ε_{SK}			ε_C			ε_S		
	Min	Max	Avr	Min	Max	Avr	Min	Max	Avr
Surface1	30.3	1.0	23.6	133.3	0.4	35.4	109.1	0.2	34.2
Surface2	0.0	3.6	6.1	50.0	101.2	11.0	20.0	42.4	11.0
Surface3	0.0	28.5	2.5	0.0	332.1	0.0	0.0	37.2	2.5

4. Conclusions

The proposed method shows much promise in terms of the accuracy, stability as well as reliability. The approach reduces the relative error up to 304% with regard to the maximum scallop and up to 12% with regard to the average scallop with the reference to conventional methods. We conjecture that the new approach can be used for efficient iso-scallop tool path generation.

5. References

[1] Hu P, Chen L and Tang K 2017 *Comput. Aided. Des.* **83** 33–50
[2] Yuwen S, Dongming G, Zhenyuan J and Haixia W 2006 *Int. J. Adv. Manuf. Technol.* **28(7-8)** 721-26
[3] Suresh K and Yang D C H 1994 *J. Eng. Ind.* **116(2)** 253–9
[4] Lin R-S and Koren Y 1996 *J. Eng. Ind.* **118(1)** 20-8
[5] Zhen-yuan J, Xiao-xuan Z, Jian-wei M, Si-yu C, Feng-ze Q and Zhao L 2019 *Precis. Eng.* **55** 217-30
[6] Guanying H, Xin J, Cheng S, Zehong L, Yuwen S, Zhiming Z and Deyi X 2019 *Int. J. Precis. Eng. Manuf.* **20** 777–90
[7] Dang L V, Vacharanukul K and Makhanov S 2019 *Int. J. Adv. Manuf. Technol.* **105** 1853–82
[8] Zhiping L, Xiongbing L and Bing Y 2019 *Int. J. Adv. Manuf. Technol.* **102(1-4)** 647-58
[9] Jinting X, Longkun X, Yuwen S, Yuan-Shin L and Jibin Z 2019 *J. Comput. Inf. Sci. Eng.* **19(4)** 041015
[10] Zhiwei L, Jianzhong F, Hongyao S and Wenfeng G 2014 *Comput. Aided. Des.* **56** 120-32
[11] Yang S H and Lee S-G 2002 *Int. J. Adv. Manuf. Technol.* **20(5)** 326-33

Acknowledgments
This research is supported by the Center of Excellence in Biomedical Engineering of Thammasat University.

IOP Conf. Series: Materials Science and Engineering **840** (2020) 012007 doi:10.1088/1757-899X/840/1/012007

Effects of Transition Zone on Magnetic Properties of Low Temperature Oxidation of Magnetite Particle: Comparison of Experiment and Micromagnetic Modeling

K Ge[1*] and W Williams[2]

[1]State Key Laboratory of Nuclear Resources and Environment, East China University of Technology, Nanchang, 330013, China
[2]School of GeoSciences, University of Edinburgh, Edinburgh EH9 3JW, UK

*E-mail: kunpeng.ge@yahoo.com

Abstract. A relationship of hysteresis parameters and oxidation of ultrafine magnetic particles on both experimental measurements and micromagnetic simulations is obtained through a step-by-step oxidation of magnetite. Numerical simulations of hysteresis loop and microstructure of a core-shell geometry with transition zone using a multi-layer structure show two categories of behaviour for magnetic grains during oxidation. First, the SD (Single Domain, <70 nm) and lager SV (Single Vortex, >130 nm) particles remain unchanged ratio of saturation remanence to saturation magnetization (M_{rs}/M_s), and slightly decreased coercivity (B_c) during oxidation. Second, the fine SV particles (80 nm to 120 nm), the hysteresis parameters respectively increase and dramatic decrease at the early and late stage of oxidation, and the micromagnetic behaviors vary significantly. Finally the hysteresis parameters of larger SV particles remain nearly unchanged during oxidation. The predicted magnetic properties for the core-shell model exhibit better agreement with experimental data than that of previously used core-shell geometry (a stoichiometric core surrounded by an oxidized shell). It indicates that the magnetic properties of partially oxidized magnetic grains are controlled by the multi-layer coupling effects and can record paleomagnetic signals.

1. Introduction

Wide practical use of magnetic core/shell nanoparticles in biomedicine, electronics, catalysis and composite materials [1–2] has given rise to numerous studies on the effects of various factors on their magnetic properties. In Geosciences, the core-shelled oxidized magnetite or titanomagnetite particles is one of the paramount important effects of weathering under natural atmospheric conditions [3]. The oxidation of magnetite or titannomagnetite starts at its surface, where Fe^{2+} is oxidized to Fe^{3+}. It is either partially removed from the crystal or it reacts with oxygen to form a new crystal layer [4]. Further oxidation is a diffusion process driven by the oxidation gradient: Fe^{2+} diffuses from the interior of the grain to the surface, leaving vacancies in the interior. Because of the isolation effect of the oxidized layer and the rapid reduction of the solid state diffusion coefficient at low temperature [5-6], a strong oxidation gradient builds up close to the grain surface: an oxidized shell is formed around an unoxidized core (so called the core-shell structure).

However, based on the diffusion process of iron cations, the actual oxidation of single magnetite crystals occurs continuously and nonlinearly from the interior of the grain to the surface [4-5]. [1] suggested a low temperature method for testifying the oxidation of magnetite and considered that a transition zone may exist between the surfacial maghemite shell and the magnetite core, where the lattice parameters gradually change. The core-shell structures of oxidized magnetite has attracted interests from researchers both experimentally and numerically [7-10]. Regardless of these attempts, how the core-shell structure, especially the transition zone affects the magnetic behaviors of the partially oxidized magnetitic particles has not been well resolved.

Micromagnetic modeling is a novel approach to explain magnetization reversal, hysteresis effects and research on complex microstructure influence, and typically efficient to present internal images of the microscopic domain structures for further evaluating how magnetic record varies under an applied field [11]. The aim of this study is to correlate the micromagnetic results with experiment measurements using a multi-layer core-shelled model during the whole process of oxidation. The more realistic core-shell model will provide a better understanding of low temperature maghemization.

2. Samples and experiments

The experiment data of synthetic reduced and oxidized magnetite were used as described by [10]. The low temperature oxidized magnetite were obtained by heating reduced magnetite to different temperatures for various periods of time. X-ray cell edges for the stoichiometric and partial oxidized magnetite particles were measured using an X-ray spectra using Cu-Kα radiation. By comparison with the corrected curve [12] of cell size and oxidation parameter z, which is determined by [13] with the following formula:

$$Fe^{2+} + (z/4)O_2 \rightarrow (1-z)Fe^{2+} + zFe^{3+} + (z/2)O^{2-} \quad (1)$$

Hysteresis loops of the dispersed powders were measured on MicroMag 3900 Vibrating Sample Magnetometer (VSM; Princeton Measurements Corp., USA) at room temperature with field ranging from -1.0 to 1.0 Tesla.

3. Micromagnetic modeling

The calculation of the core-shell structure was performed using a finite element method (FEM). Micromagnetic models calculate stable magnetic structures by minimizing the total magnetic energy E_{total} as the initial guess, which is the sum of the exchange energy E_{ex}, the magnetostatic energy E_d, the anisotropy E_{anis} [14] and the external field energy E_h. E_d is calculated by using the fast-Fourier transforms (FFT) method to accelerate the computation [15]. That is to say:

$$E_{total} = E_{ex} + E_{anis} + E_d + E_h \quad (2)$$

In the micromagnetic modeling, the normal material parameters appropriate to magnetite at room temperature were used, namley Exchange Constant ($A_{ex} = 1.33 \times 10^{-11} Jm^{-1}$), Magnetocrystalline Anisotropy ($K_1 = -1.24 \times 10^4 Jm^{-1}$) and Satruration Magnetization ($M_s = 4.8 \times 10^5 Am^{-1}$) [16-17]. For maghemite, the corresponding principle magnetic parameters applied in this core-shell model are $1 \times 10^{-11} Jm^{-1}$, $-4.6 \times 10^3 Jm^{-1}$, and $3.8 \times 10^5 Am^{-1}$ respectively [18]. [5] described a kinetics numeric modeled function of depth with iron diffusion (oxidation state) for continuous oxidized magnetite grains (Figure 1). The material parameters of the partially oxidized layer for the core-shell model were hence determined by simply linearly interpolations of the parameters of magnetite and maghemite on the basis of their oxidation parameter. Averaged exchange parameters between bordered layer were employed on the boundaries of layers [10].

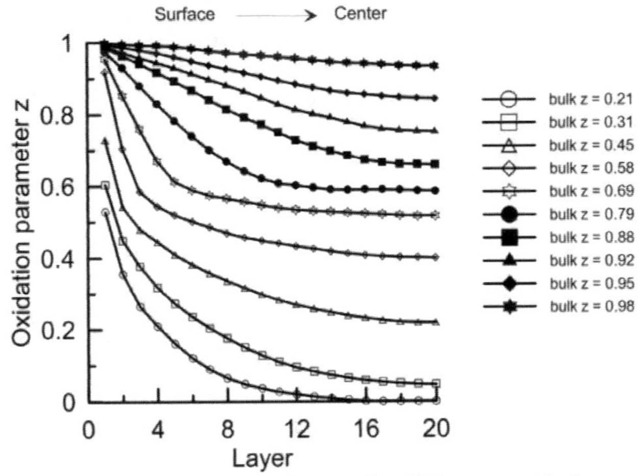

Figure 1. Oxidation parameter z as a function of layer for different oxidation stage of bulk magnetite grain (Modified from [5]).

Simulations of hysteresis were conducted with applied field varying from 180 mT to -180 mT in steps of 5 mT. Values for the coercive force (B_c) and ratio of saturation remanence to saturation magnetization (M_{rs}/M_s) are averaged from the modeling results with applied fields aligned along the easy <111>, hard <100> and intermediate <110> directions.

4. Results and discussions

The coercivity and the ratio of saturation remanence to saturation magnetization as a function of oxidation parameter are shown in Figure 2. For grains from 40 nm to 70 nm, the B_c decrease gradually with the proceeding of oxidation process. While the M_{rs}/M_sstays stable with the increase of oxidation parameter. The hysteresis behaviors offine SV grains (80 to 120 nm) show a greater variability as a function of oxidation. It appears that hysteresis parameters of these grains during oxidation are larger than those of two end members, and specifically, dramatic increase and decreases could be seen at the early and late stage of oxidation, respectively. For larger grains (140 to 160 nm), the parameters remain nearly unchanged during oxidation in comparison of finer SV particles. Specifically, the B_c decrease slightly with the increase of oxidation parameter, however the M_{rs}/M_s shows slight increase with the oxidation proceeds.

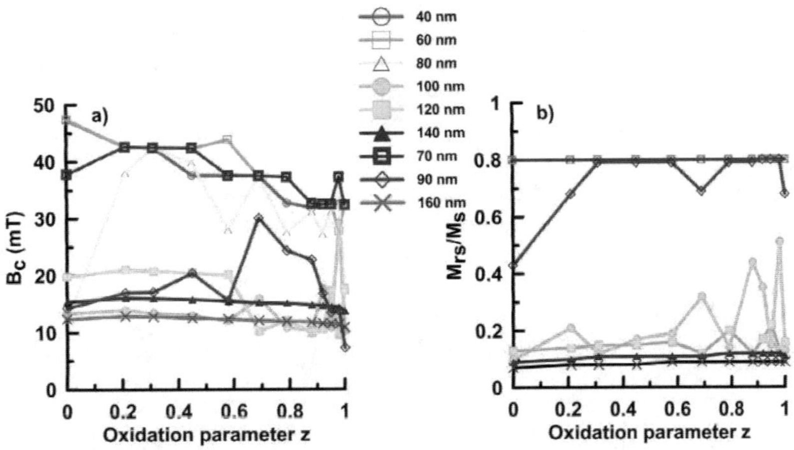

Figure 2. Micromagnetic results of (a) coercive field (B_c) and (b) ratio of saturation remanence to saturation magnetization (M_{rs}/M_s) versus oxidation parameter z for multi-layer core-shelled models.

As seen from Figure 3, the hysteresis parameters of experiment data as a function of oxidation states exhibit similar behaviors, which is a slight increase increasing up to z = ~0.9, and then decreases dramatically. A dramatic decrease of hysteresis parameters appears both for multi-layer core-shell model and 2-layer core-shell model before magnetite is totally oxidized, which is well consistent with the experimental observations. The micromagnetic data for multi-layer core-shelled model show that both trend and values of B_c and M_{rs}/M_s versus oxidation parameters are more similar to the experimental observation thanthe data of the simple no-transitioned core-shell model. In particular, the M_{rs}/M_s as a function of oxidation parameter is almost indistinguishable with the experiment results.

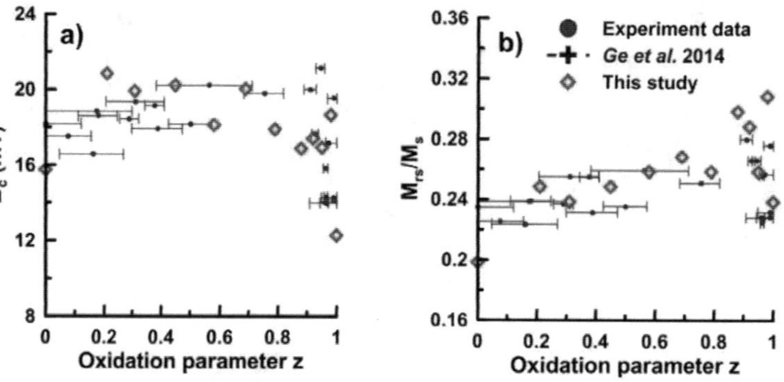

Figure 3. Weighted micromagnetic modeling results of (a) coercive field and (b) ratio of saturation remanence to saturation magnetization, versus oxidation parameter z, according to the grain size distribution, in comparison with previous experiments by [10]. Note that the interaction effects have been considered in the weighted multi-layer model.

Note that a slight increase of coercive force shows up in the micromagnetic model at an early oxidation stage (Figure 3a).There are experimental explanations of this increase of remanence stability. For instance, [19] proposed that the increase of coercivity is believed to result from stress in the absence of any other quantitatively plausible mechanism. The followed decrease may be due to a decrease in the magnetostriction constant as a function of oxidation. However, this slight change probably results from the multi-layer coupling effect during oxidation.

5. Conclusions

Based on experimental characters of low temperature oxidized magnetic particles, micromagnetic simulations were performed to investigate the hysteresis parameters versus oxidation state as a comparison. Using the FEM calculation of a multi-layer structure, the averaged results based on grain size distribution are in good agreement with the experimental data, and show two types of variations of magnetite properties from SD to SV particles. The dramatic change of hysteresis parameters before totally oxidized in experiments is mainly because of the variability of magnetic properties for partially oxidized fine SV grains. Magnetic particles of other sizes are magnetically stable during oxidation. Overall, the partially oxidized magnetite prefers an multi-layer coupled core-shell structure, and is capable of record reliable magnetic signals.

6. References

[1] Afremov L, Anisimov S, and Iliushin I 2018 Size effect on the hysteresis characteristics of a system of interacting core/shell nanoparticles *Journal of Magnetism and Magnetic Materials* **447** 88–95

[2] Kim Y, Park K Y, Dong M J, Yun M S, Han S K, Yong J C, Myung Y, and Park J 2010 Synthesis of Au–Cu2S Core–Shell *Nanocrystals and Their Photocatalytic and Electrocatalytic Activity* **114** 22141–22146

[3] Özdemir O and Dunlop D J 2010 Hallmarks of maghemitization in low-temperature remanence cycling of partially oxidized magnetite nanoparticles *Journal of Geophysical Research-Solid Earth* **115** 1–10
[4] O'Reilly W 1984 *Rock and mineral magnetism* (Glasgow: Blackie)
[5] Gallagher K 1968 Mechanism of oxidation of magnetite to gamma-Fe_2O_3 *Nature* **217** 1118–1121
[6] Askill J 1970 *Tracer Diffusion Data for Metals, Alloys, and Simple Oxides* (New York: IFI Plenum)
[7] Cui Y, Verosub K L, and Roberts A P 1994 The effect of low-temperature oxidation on large multi-domain magnetite *Geophysical research letters* **21** 757–760
[8] van Velzen A J and Dekkers M J 1999 Low-temperature oxidation of magnetite in loess-paleosol sequences: a correction of rock magnetic parameters *Studia geophysica et geodaetica* **43** 357–375
[9] Liu Q S, Banerjee S K, Jackson M J, Deng C L, Pan Y X, and Zhu R X 2004 New insights into partial oxidation model of magnetites and thermal alteration of magnetic mineralogy of the Chinese loess in air *Geophysical Journal International* **158** 506–514
[10] Ge K, Williams W, Liu Q, and Yu Y 2014 Effects of the core–shell structure on the magnetic properties of partially oxidized magnetite grains: experimental and micromagnetic investigations *Geochemistry, Geophysics, Geosystems* **15** 2021–2038
[11] Fidler J and Schrefl T 2000 Micromagnetic modelling-the current state of the art *Journal of Physics D: Applied Physics* **33** R135–R156
[12] Readman P and O'reilly W 1972 Magnetic properties of oxidized (cation-deficient) titanomagnetites $(Fe, Ti, \square)_3O_4$ *Journal of Geomagnetism and Geoelectricity* **24** 69–90
[13] Readman P and O'Reilly W 1971 Oxidation processes in titanomagnetites *Zeitshrift für Geophysik* **37** 329–338
[14] Brown W F 1963 *Micromagnetics* (New York: Interscience)
[15] Wright T, Williams W, and Dunlop D 1997 An improved algorithm for micromagnetics *Journal of Geophysical Research* **102** 12085–12094
[16] Pauthenet R and Bochirol L 1951 Aimantation spontanée des ferrites *Journal de Physique et le Radium* **12** 249–251
[17] Heider F and Williams W 1988 Note on Temperature-Dependence of Exchange Constant in Magnetite *Geophysical Research Letters* **15** 184–187
[18] Dunlop D J and Özdemir Ö 2001 *Rock magnetism: fundamentals and frontiers* (Cambridge: Cambridge University Press)
[19] Özdemir Ö and O'Reilly W 1982 Magnetic hysteresis properties of synthetic monodomain titanomaghemites *Earth and Planetary Science Letters* **57** 437–447

Acknowledgments
This work was supported by the National Natural Science Foundation of China (Grant No. 41964003, and 41964006). W. W. also acknowledge Natural Environment Research Council (UK) grants NE/S001018 and NE/S01197.

IOP Conf. Series: Materials Science and Engineering **840** (2020) 012008 doi:10.1088/1757-899X/840/1/012008

Reflection spectra of a thin-film GeSbTe diffraction grating on a silicon nitride waveguide

R T Minnullin[1,2,*], A A Sapegin[1,2], M E Makarov[1,3], D S Korolev[1,4] and M Yu Barabanenkov[1,3]

[1] JSC Molecular Electronics Research Institute, Moscow, Zelenograd, 124460, Russia
[2] Moscow Institute of Physics and Technology, Dolgoprudny, 141701, Russia
[3] Institute of Microelectronics Technology RAS, Chernogolovka, 142432, Russia
[4] Lobachevsky State University of Nizhny Novgorod, Nizhny Novgorod, 603950, Russia

[*] Email: ramil.minnullin@phystech.edu

Abstract. Recent advances in nanophotonics are due to the implication of new approaches to the photonic devices and components design, not only related to structural features, such as subwavelength periodic arrangements, but also new materials, e.g., phase-change materials like GeSbTe (GST) alloys. We consider recently proposed optical non-volatile GST memory cell with a GST diffraction grating instead of a continuous film placed on a silicon nitride waveguide. The grating allows diminishing the energy budget of an incident electromagnetic beam in case of an optically induced phase transition of GST due to excitation of the resonant guided mode in the grating. The excitation of this mode results in anomalous reflectance spectra of the waveguide-grating structure. Here, we present the reflection spectra of GST diffraction gratings on a silicon nitride waveguide calculated with the use of the matrix Riccati equation technique in the theory of multiple electromagnetic wave scattering in inhomogeneous media. We show how the reflection changes with variation of different parameters – grating period and height, incident wave polarization, and phase of the GST film.

1. Introduction

Nanophotonics is a rapidly growing interdisciplinary research field [1] that emerged as an alternative to modern nanoelectronics, which faces certain problems in progressing through the scaling of standard planar CMOS technology [2]. Nanophotonics provides exceptional opportunities for developing high-performance devices for optical telecommunications, computing, quantum cryptography, as well as for biological and medical applications [3]. In particular, it is anticipated that the integrated photonics and phase-change materials (PCMs) [4, 5] enable the development of a photonic non-volatile memory [6], which is essential to overcome the so-called von-Neumann bottleneck meaning the information traffic jam between the processor and the memory.

To store data, PCM-based electronic random access memory (PRAM or PCRAM [4, 6]) exploit the large contrast in electrical resistance between the amorphous state and the crystalline state of PCMs [4]. Typical PCM materials, such as chalcogenide glass compounds (GeSbTe, AgInSbTe etc.), are characterized by fast and reversible phase transitions between the two above-mentioned states at elevated temperatures. Thus, PRAM utilizes only the electrical properties of PCMs, and being an electronic component, it possesses the same disadvantages as all electronic devices compared to photonic ones, namely significant RC-delays, Joule heating, and low resistance to external electromagnetic (EM) radiation.

PCMs also demonstrate a large difference in a refractive index upon a phase transition, an effect that has led to their commercial use in optical storage disks [7]. The concept of memory operation in optical disks relies on the far-field optics, which makes downscaling of such memory cells a challenging issue that requires the overcoming of the diffraction limit. On the other hand, photonic PCM-based non-volatile memory is operated in the optical near field, meaning that the sizes of memory cells are not restricted by the diffraction limit and can be reduced to nanoscales. Therefore, it allows avoiding the above-mentioned drawbacks of electronic devices.

Figure 1 illustrates the concept of a photonic memory cell proposed in [6]. A nanoscale GST film is placed directly on top of the waveguide. The guided mode is evanescently coupled into the GST film. Depending on the phase of the GST, the mode energy is absorbed inside the film to less (in the amorphous state) or greater (in the crystalline state) extent. Accordingly, the signal at the waveguide output weakens slightly in the first case and significantly in the second, thus, defining the levels of logical '1' and '0'. The state of a GST film can be changed with the use of a more intense light pulse needed to heat the GST to the transition temperatures, providing 'Write' and 'Erase' operations. It is further reported that even intermediate crystallization/amorphization states of a GST film can be achieved, allowing for a multilevel cell that can be used as a memristor for neural network and artificial intelligence applications.

Figure 1. Schematic illustration of the photonic non-volatile memory cell with GST [6].

In this report, we propose a GST film in the form of a strip-like diffraction grating (figure 2) and numerically investigate the reflection of an EM wave from a waveguide-grating structure.

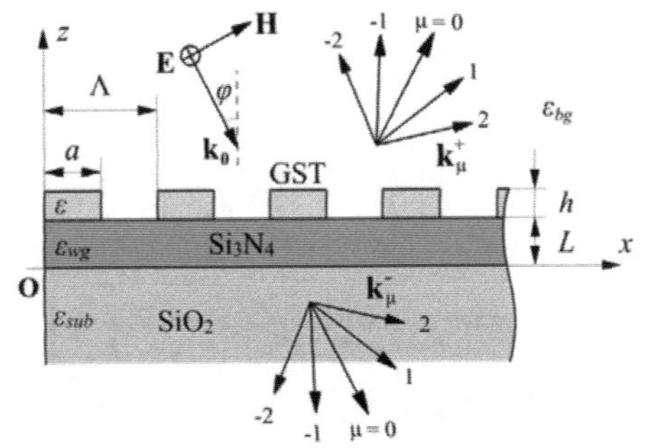

Figure 2. Schematic view of an *x-z* plane of the considered structure: the Si_3N_4 waveguide on the SiO_2 substrate with a rectangular GST diffraction grating (strip height h, width a, period Λ). The incident plane wave is represented with wavevector $\mathbf{k_0}$ and **E**- and **H**-field vectors (**E**-field vector parallel to the grating strips denotes the TE polarization). Vectors $\mathbf{k}_\mu^+$ indicate angular orders of the reflected EM radiation, $\mathbf{k}_\mu^-$ – angular orders of the radiation, transmitted to the substrate (see Equations (1, 2)).

Using a grating instead of a continuous film has a three-fold aim. First, it allows diminishing the energy budget of an incident EM beam in the case of an optically induced phase transition of GST. Second, the evanescent part of the guided mode could be more strongly coupled to the grating. Third, the phase state of PCM may be simply verified on the basis of magnitudes of angular orders in the reflection spectra of relatively weak external EM radiation.

Indeed, it is well known that under certain conditions, a resonance phenomenon occurs in waveguide-grating structures comprised of a substrate or a thick buried oxide layer (SiO_2 in figure 2), a thin waveguide layer (Si_3N_4), and a layer in which a grating is fabricated (GST). Hessel and Oliner demonstrated the resonant excitation of surface waves in metallic grating structures [8]. Neviere generalized this picture to deal with structures that include gratings and dielectric waveguides [9]. Later on, it was confirmed both numerically [10, 11] and experimentally [12] that the excitation of a guided mode in the waveguide and even in the grating layer itself [13] exhibits anomalous (maxima) zeroth-order reflections in the reflection spectra of the waveguide-grating structures. The latter case [13] is the most interesting in terms of the PCMs phase state because the above-mentioned resonant excitation of waves in the grating causes an increase in the incident EM wave energy absorption in the PCM grating strips.

2. Matrix Riccati equation method

We calculate the reflection spectra and the EM field spatial distributions for the waveguide-grating structure shown in figure 2 applying an analytical approach of "transfer relations" and "invariant embedding" method to the theory of EM wave multiple scattering in inhomogeneous media [14, 15]. This method uses the Riccati type equation for the matrix wave reflection coefficient and the associated differential equation for the matrix wave transmission coefficient. It allows considering the effects of strong energy transformations between homogeneous and evanescent waves in the near wave zone of the grating, which appear due to the subwavelength or comparable to the wavelength dimensions of the considered structures.

Let us assume that a plane TE polarized EM wave with wavelength λ is incident from a background dielectric medium, $\varepsilon_{bg} = 1$, on a periodic structure with period Λ, height h and dielectric permittivity $\varepsilon(\mathbf{r}, \omega)$ as a function of coordinates $\mathbf{r}$ and incident wave frequency ω, at an angle of incidence φ (see figure 2). The method of the transfer relations allows representing the electric field of reflected and transmitted radiation as a superposition of the plane waves with certain wave vectors ($\mathbf{k}_\mu^\pm$) and certain coefficients ($R_{\mu 0}$, $T_{\mu 0}$):

$$E_y(r) = \begin{cases} \exp\left(i\mathbf{k}_0\left(\mathbf{r} - h\mathbf{n}_z\right)\right) + \sum\limits_{\mu=-\infty}^{\infty} R_{\mu 0}(h)\exp\left(i\mathbf{k}_\mu^+\left(\mathbf{r} - h\mathbf{n}_z\right)\right), & z > h \\ \sum\limits_{\mu=-\infty}^{\infty} T_{\mu 0}(h)\exp\left(i\mathbf{k}_\mu^-\mathbf{r}\right), & z < 0 \end{cases}, \tag{1}$$

$$\mathbf{k}_\mu^\pm = \begin{pmatrix} \beta_\mu \\ 0 \\ \pm\sigma_\mu \end{pmatrix} = \begin{pmatrix} k_0 \sin\varphi + 2\pi\mu/\Lambda \\ 0 \\ \pm\sqrt{k_0^2 - \beta_\mu^2} \end{pmatrix}. \tag{2}$$

The reflection and transmission coefficients are derived from the solution of the following matrix differential equations:

$$\mathrm{R}_z' = \mathrm{R}\mathrm{A}(z)\mathrm{R} + \left(\mathrm{A}(z) + \mathrm{C}\right)\mathrm{R} + \mathrm{R}\left(\mathrm{A}(z) + \mathrm{C}\right) + \mathrm{A}(z), \tag{3}$$

$$\mathrm{T}_z' = \mathrm{T}\left(\mathrm{A}(z) + \mathrm{C} + \mathrm{A}(z)\mathrm{R}(z)\right), \tag{4}$$

$$\mathrm{R}, \mathrm{T}\big|_{z\to 0} = \mathrm{R}_0, \mathrm{T}_0, \tag{5}$$

where R, T, A, and C are $(2n+1)\times(2n+1)$ -dimensional matrices with complex values and n is selected in a special way to achieve convergence. The components of the matrices A and C are defined as follows:

$$A_{\mu\nu}(z) = i\frac{2\pi}{\lambda^2\sigma_\mu} f_{\mu-\nu}(z), \tag{6}$$

$$C_{\mu\nu} = i\sigma_\mu\delta_{\mu\nu}, \tag{7}$$

and $f_\alpha(z) = \frac{\pi}{\Lambda}\int\limits_0^\Lambda \varepsilon(x,z)\exp\left(-i\frac{2\pi\alpha}{\Lambda}x\right)\mathrm{d}x$ is a transformation function, it describes the interaction between propagating and evanescent waves at multiple scattering in periodical inhomogeneous media.

For TM polarization equations (3)–(6) and transformation function slightly differ. A more detailed description of the method is provided in [15].

3. Results and discussion

The reflection spectra (figure 3) of the GST diffraction grating on a 220 nm thick Si_3N_4 layer and SiO_2 substrate and spatial EM field distributions (figure 4) are computed for 1550 nm wavelength at normal incidence ($\varphi = 0$) and different parameters (the grating height and the period, the phase state of the GST material, TE and TM polarizations of the incident light). The grating period is varied from 0.2λ to 10λ (from 0.31 to 15.5 μm), height – from 5 to 70 nm, grating duty cycle (a/Λ) is maintained 0.5. The refractive indices of SiO_2 and Si_3N_4 at a given wavelength are 1.466 and 1.996, respectively. The refractive indices and the absorption coefficients of GST in the amorphous and crystalline states are $n_a = 4.4$, $k_a = 0.2$ and $n_c = 7.2$, $k_c = 1.4$, respectively, at the same wavelength [16].

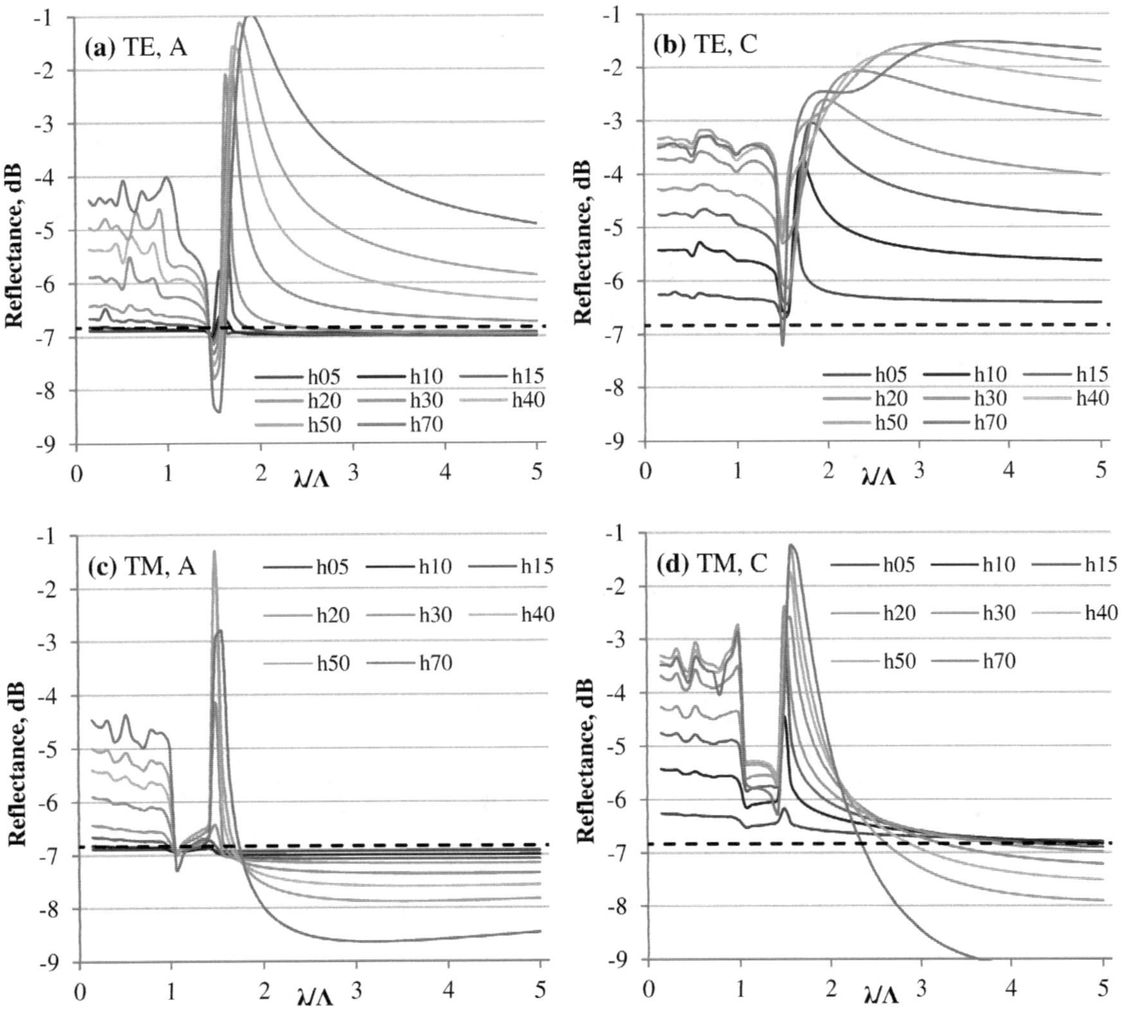

Figure 3(a, b, c, d). Reflection spectra of a thin-film GST diffraction grating on Si_3N_4 waveguide on SiO_2 substrate in the case of TE (a, b) and TM (c, d) polarizations of a normally incident plane wave for the amorphous (left side – panels a, c) and crystalline (right side – panels b, d) phases of GST at different grating heights (h in nm). The dashed horizontal line indicates a "zeroth" reflection level of the Si_3N_4 layer on SiO_2 substrate without a grating.

Each graph in figure 3 shows a common sharp decline at a point $\lambda/\Lambda = 1.466$, which defines the (lower) boundary of the "subwavelength" regime of a grating [17]. Interestingly, this sharp decline is typically followed by a more or less sharp increase in the reflectance (see figure 3a, 3c and 3d). However, the grating in the crystalline phase in the case of TE polarization (figure 3b) demonstrates a substantially different behavior resulting in at least three types of curves: first, lines with steep growth and a corresponding decrease (heights from 5 to 15 nm); second, lines with steep growth and a gradual decrease (heights from 20 to 30 nm); and third, lines with a sharp increase and an extra inflection followed by even a more gentle decrease (heights 40, 50 and 70 nm).

Another notable point is $\lambda/\Lambda = 1.0$ that defines the first Wood–Rayleigh grating resonance [18], also known as a "mode opening" and physically meaning the transformation of ±1 order modes from evanescent to propagating. According to the graphs in figure 3c and 3d, TM polarization allows for a clearer display of the Wood–Rayleigh resonances and sharper regime-transition peaks, forming a gap with a low reflectance level between these two points ($\lambda/\Lambda = 1.0$ and $\lambda/\Lambda = 1.46$).

Figure 4 demonstrates spatial distributions of electric and magnetic fields of the reflected and transmitted to the substrate radiation. The field distributions are calculated at extraordinary points.

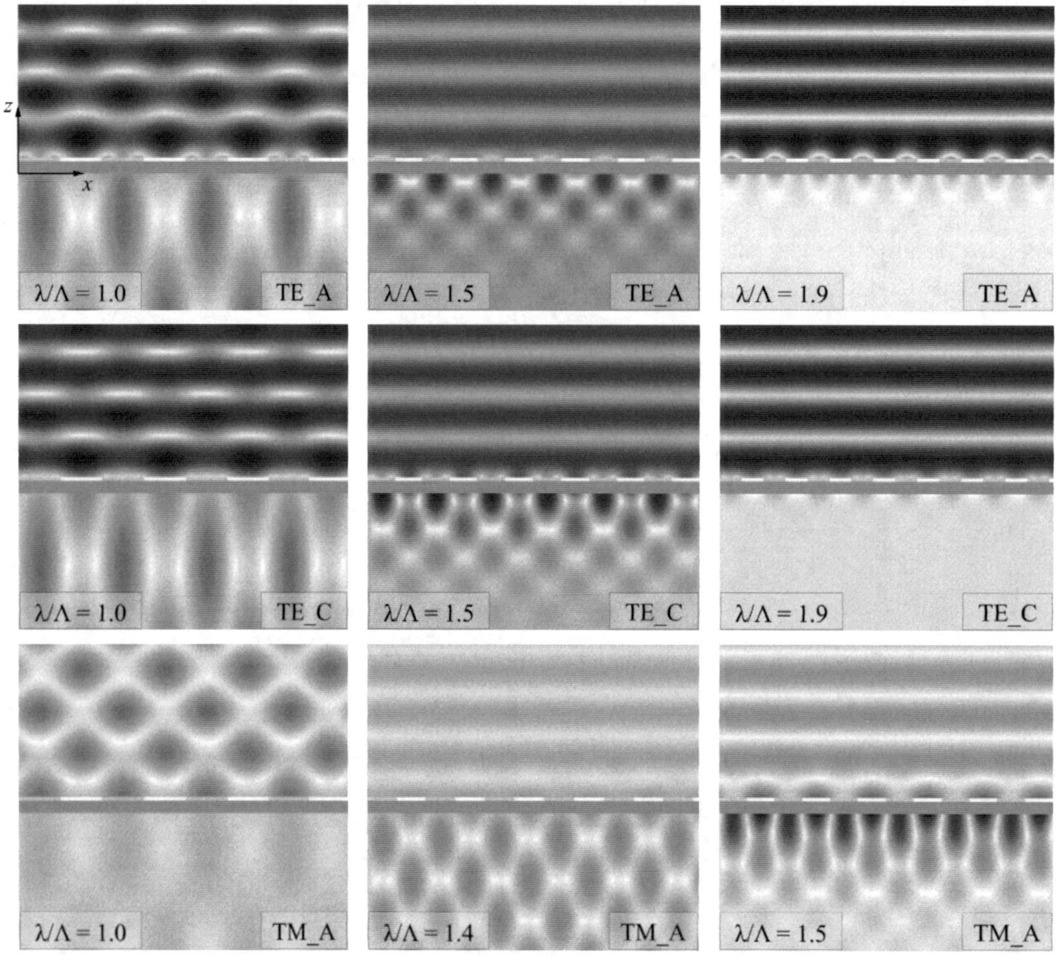

Figure 4. Spatial distributions of the electric field (TE polarization case) and the magnetic field (TM polarization) of the reflected and transmitted to the substrate radiation for gratings at different λ/Λ ratios and GST phases (A – amorphous, C – crystalline). The grating height is 70 nm. The deeper color corresponds to a more intense field.

Point $\lambda/\Lambda = 1.0$ corresponds to the first-order Wood–Rayleigh grating resonance; point $\lambda/\Lambda = 1.5$ (and 1.4 in TM case) corresponds to the edge of subwavelength regime (minimum of reflectance); and point $\lambda/\Lambda = 1.9$ corresponds to maximum of reflectance. The dependencies for TM polarization in figure 3(c, d) are similar for both phase states of the GST grating, and the magnetic field distributions are also similar. For this reason, figure 4 provides field pictures only in the cases of the amorphous GST state for TM polarization. The electric field distributions for TE polarization in the amorphous and crystalline states of GST (figure 4, rows 1 and 2) also have similar patterns despite significantly different curve behaviors (see figure 3a, b).

There are several ways to interpret the results. If the diffraction grating is used for optical switching of the GST phase by external irradiation, the reflection spectra help identify the parameters when high or resonant absorption in GST occurs, and the field distributions provide a visual representation of the regions of energy concentration. Furthermore, the diffraction grating may be even used as an input port of a memory cell exploiting different levels of coupling efficiency that strongly depend on reflection levels (see, for example, figure 3a, b). Another variant is possible when the diffraction grating is used for enhancing the existing optical phase-switching mechanism by reducing the time and energy parameters via concentrating the EM field in certain regions of the grating.

In conclusion, we demonstrate one possible way of applying a structural design approach involving diffraction gratings with subwavelength or compared to the wavelength dimensions to the concept of a photonic non-volatile memory cell that utilizes the GST material with a switchable phase. This can be particularly helpful for designing memory cells with different mechanisms of the GST phase switching. Furthermore, the same approach may be used for designing multilevel memory cells, when, for example, the phases of each constituent element of the GST grating is switched separately. To this end, waveguide-grating structures may be considered as a conventional continuous waveguide and a discrete waveguide composed of a linear chain of nanoparticles [19]. In this case, the problem is an EM coupling between the modes of a continuous waveguide and the collective modes of a nanoparticle chain.

4. References

[1] Fabrizio E D and Stockman M 2019 *Nanophotonics* **8** 1443
[2] Krasnikov G Ya and Zaitcev N A 2009 *Nano and Microsystem Technique* **1** 2 (*in Russian*)
[3] Cheben P et al. 2018 *Nature* **560** 565
[4] Zhang W et al. 2019 *Nature Reviews Materials* **4** 150
[5] Yamada N et al. 1991 *J. Appl. Phys.* **69** 2849
[6] Ríos C et al. 2015 *Nature Photon.* **9** 725–732
[7] Ohta T et al. 1998 *IEEE Trans. Magn.* **34** 426
[8] Hessel A, Oliner A A 1965 *Appl. Opt.* **4** 1275–97
[9] Neviere M 1980 *Electromagnetic Theory of Gratings* ed. R. Petit (Berlin: Springer-Verlag)
[10] Popov E et al. 1986 *Optica Acta* **32** 607–629
[11] Golubenko G A et al. 1985 *Sov. J. Quantum Electron.* **15** 886
[12] Avrutsky I A et al. 1989 *J. Modern Opt.* **36** 1303
[13] Wang S S et al. 1990 *J. Opt. Soc. Amer.* A 7 1470
[14] Barabanenkov Yu N and Barabanenkov M Yu 1999 *Electromagnetic waves* **24** *PIERS* 39
[15] Barabanenkov M Yu and Barabanenkov Yu N 2006 *J. Opt. Soc. Amer.* A **23** 581
[16] Strand D et al. 2006 EPCOS2006 (European Phase Change and Ovonics Symposium)
[17] Halir R et al. 2015 *Laser Photonics Rev.* **9** 25
[18] Maystre D 2012 *Theory of Wood's anomalies,* in *Plasmonics. From Basics to Advanced Topics.* ed. S Enoch, N Bonod (Berlin: Springer), 39
[19] Barabanenkov Yu N and Barabanenkov M Yu 2017 *J. Opt. Soc. Amer.* A **34** 321

Acknowledgements
This study was supported by the Russian Foundation for Basic Research, project No.19-29-03040. We would also like to thank Dr Alexander Italyantscv for his support and encouragement and Dr Elena Bazanova for assistance with academic English writing.

IOP Conf. Series: Materials Science and Engineering **840** (2020) 012009 doi:10.1088/1757-899X/840/1/012009

Effects of Surface in the IR and Raman Spectrum of Porous Silicon Carbide

R Bermeo, L. Arellano, A Trejo, F Salazar, M. Calvino, A Miranda, and M Cruz-Irisson

Instituto Politécnico Nacional, ESIME Culhuacan, Av. Santa Ana 1000, C. P. 04440, Ciudad de México, México

Email: alejandtb13@gmail.com

Abstract. Porous Silicon carbide has been identified as an attractive material for its use as electrode in supercapacitors, however the theoretical investigations about its properties, specially its vibrational properties, are still scarce. In this work the effect of the Si-C surface ratio on the vibrational properties, IR and Raman spectrum of porous silicon carbide was studied using the first principles density functional perturbation theory. The porous structures were modelled by removing atoms in the [001] direction from an otherwise perfect SiC crystal using the supercell scheme. The morphology of the pores was chosen so there would be more Si or C in the pore surface. The results show that the vibrational properties, and thus the IR and Raman spectrum of the porous SiC change depending if the pore surface is either Si or C rich, having the Si-rich pores more low frequency modes due to its higher mass. Also, the effects of phonon confinement are lessened by the effect of surface passivation, thus indicating that the surface plays an important role in the IR and Raman characterization of these structures.

1. Introduction

In recent years, the development of alternative energy storage devices has been the focus of many theoretical and experimental investigations. Among the most attractive devices for energy storage are the supercapacitors which, as their name indicates, are capacitors that far surpass the capacity of regular capacitors, being attractive due to its fast charge delivery and longer Lifecyle compared to batteries, making it suitable for multiple applications such as in electric vehicles [1] . However, in order to improve the capacity and lifecycle of this devices, multiple anode materials have been proposed, where porous Silicon Carbide (pSiC) is one of the most attractive due to its high surface area and high electron mobility [2, 3]. Additional applications of pSiC have also been identified, such as in Li ion batteries [4], Biological sensors [5], H sensors[6], amongst others. Nevertheless, the properties of pSiC have been seldom theoretically studied [7, 8], specially its vibrational properties where only a few works exist [9]. While, the effect of the surface on the vibrational properties of porous silicon carbide has been studied previously, only extreme surface of 100% Si or C were analyzed, and the effect of a mixed surface (which would be closer to the experimental ones) was not addressed [9]. Hence in this work we study the vibrational properties of porous silicon carbide with a ratio between the Si and C in the surface that allows a higher concentration of Si or C in a mixed surface. Results show that both structures are stable, and there are remarkable differences between their IR and Raman spectrums having some peaks prevalence over others depending upon the richness of Si or C in the pore surface. These results could be of great importance in the eventual development of pSiC anodes for supercapacitors and its characterization through their Raman and IR spectrums.

2. Calculation scheme

The vibrational properties of the porous structures were studied using the first principles density functional perturbation theory [10], within the generalized gradient approximation using the PBE functional [11] and norm conserving pseudopotentials [12] with 2x2x6 Brillouin zone integration grids according to the Monkhorst-Pack [13] scheme. The atomic positions and cell dimensions were optimized until the internal forces were less than 0.001 eV/Å. The porous structures were modelled by removing columns of atoms in the [001] direction from an otherwise perfect SiC crystal according to the supercell technique[14, 15], so the surface of the pore has either more Si or more C, hence a ratio between the Si and C in the pores surface was defined as $R_0=Si_{surf}/C_{surf}$. The modelled porous structures have $R_0=2$ and $R_0=0.5$ as seen in Fig 1a-b). All surface dangling bonds were passivated with H atoms since it has been observed experimentally [16] and allows for a better relaxation of the geometry with a lesser K-point sampling.

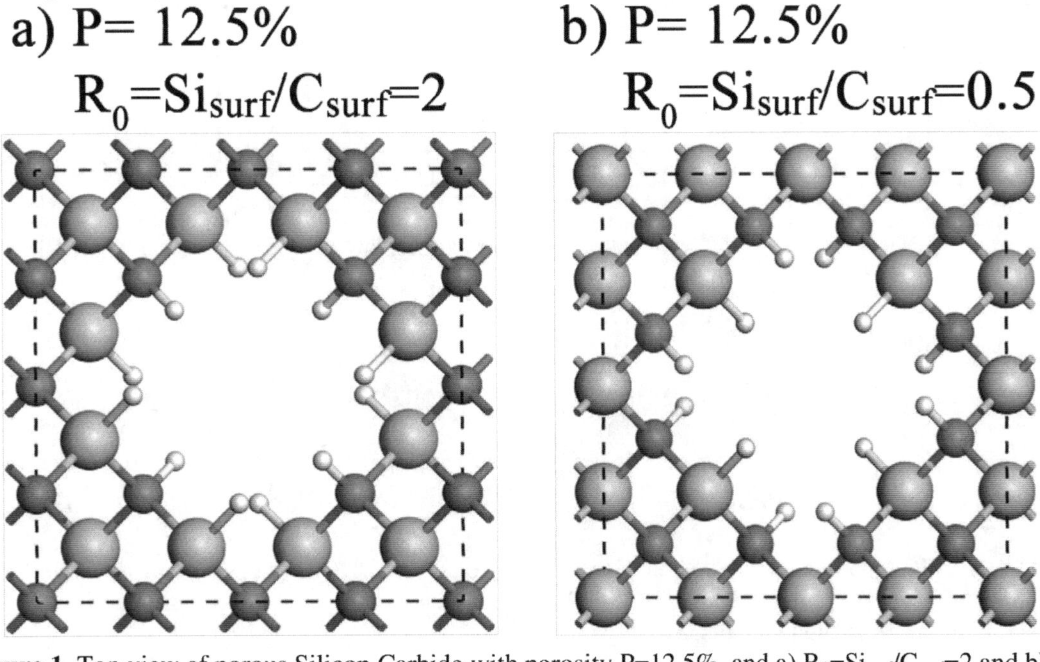

Figure 1. Top view of porous Silicon Carbide with porosity P=12.5%, and a) $R_0=Si_{surf}/C_{surf}=2$ and b) $R_0=Si_{surf}/C_{surf}=0.5$. The yellow, grey and white spheres represent, Silicon, Carbon and H atoms, respectively.

3. Results and discussion

In figure 2 it can be seen that the $R_0=2$ and $R_0=0.5$ structures have similar features in their phonon band structures and density of states, however there are also notable differences. For instance in the low frequency region (0-1200cm^{-1}), for the $R_0=0.5$ structure there is a notable gap around 600cm^{-1} which is absent in the other structure, and a lesser one around 900cm^{-1}, which is reflected in a sharper feature of the DOS since the phonon modes seem to be more packed in intervals for $R_0=0.5$ while for the other structure are spread more evenly in the low frequency region albeit a small gap around 900cm^{-1}. Also for the $R_0=2$ there is a clear splitting of the Si-H vibrational modes (frequencies between 2000 and 2500cm^{-1}) owning to the larger number of Si which could create more asymmetry in the bond stretching compared to the other structure where there is only four Si on the surface, which allows fewer different modes and hence fewer symmetric or asymmetric bond stretching in the surface. A similar behaviour can be observed for the C-H bond stretching region (around 2900cm^{-1}) where, there is a slight splitting of the bands for the R_0=0.5 case, owning to the larger number of C in the pore surface, however this splitting is less noticeable due to the lower mass of the C. It is worth mentioning that although the difference between the number of Si or C in the pore surface could be considered small, the differences in their vibrational spectrum, specially at high frequency are remarkable, which could be useful for the characterization of this material, also, the absence of negative modes in the phonon band structure and density of states indicates that the structures are stable, hence the predicted surfaces could be present in experimental samples.

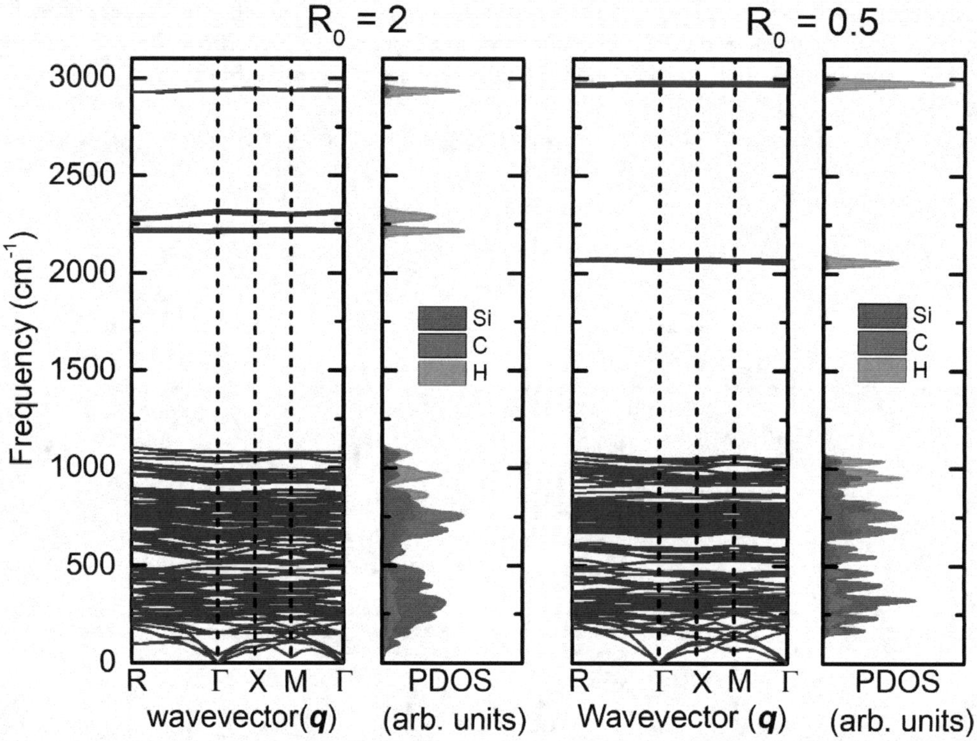

Figure 2. Phonon band structure and partial density of states (PDOS) of porous Silicon Carbide with porosity P=12.5%, and a) R_0=2 and b) R_0=0.5. The green, red and grey areas represent the Si, C and H contributions to the DOS, respectively.

Although the differences between the Phonon band structures and density of states were notable, the most noticeable changes are observed in the IR and Raman spectrums, as seen in Fig 3. For the IR case it can be seen that, although both systems show similar high intensity peaks around 1000 cm^{-1} the quantity and frequency interval of activity is different, with R_0=2 having only one notable peak around 700cm^{-1}, while, for the other case there are more notable peaks, specially the one around 1000cm^{-1}. This could be due to the presence of the passivating H which creates shifts of the dipole moment, for the R_0=2 these shifts at 1000cm^{-1} are damped due to the low concentration of C, while for the other case the higher C count allows more bond bending vibrations in this frequency interval resulting in more IR peaks. In the bond stretching vibrations something counterintuitive happens; there is a high intensity peak for the Si-H vibrations in the R_0=0.5 case while this peak is lower for the R_0=2 case, this could be due to the effect of the Si placement in the pore surface which allows a cancelation of some of the dipole moments shifts due to the proximity of the Si-H bonds[17] , something that does not happens in the R_0=0.5 case where the Si-H bonds are spread through the pore surface, allowing unperturbed shifts of the dipole moment. As for the 3000cm^{-1} modes they are almost non-existent in the IR spectrum, probably due to the negligible difference of electronegativity between C-H which induces an almost non-existent shift of the dipole moment for bond stretching vibrations since the C would not donate part of its charge that gained through the C-H bond to the C, being a bond with less polar characteristics. As for the Raman spectrum the low frequency region has lower intensity peaks compared to the stretching regions, which could be due to the low change in the polarizability of the Si-C bonds upon vibrations, while for the stretching vibrations the proximity of the H to either Si or C, although no charge is exchanged, the shape of the electron density around the bond changes, inducing a change of the polarizability which is reflected in higher Raman intensities. The expected shift of the higher optical modes of the low frequency regions due to phonon confinement is also almost non-existent due to the masking of the Si-H or C-H bonds in the surface. Also, it is worth mentioning the stark difference between the IR and Raman results which indicate that both spectroscopies are necessaries to a complete characterization of porous silicon carbide

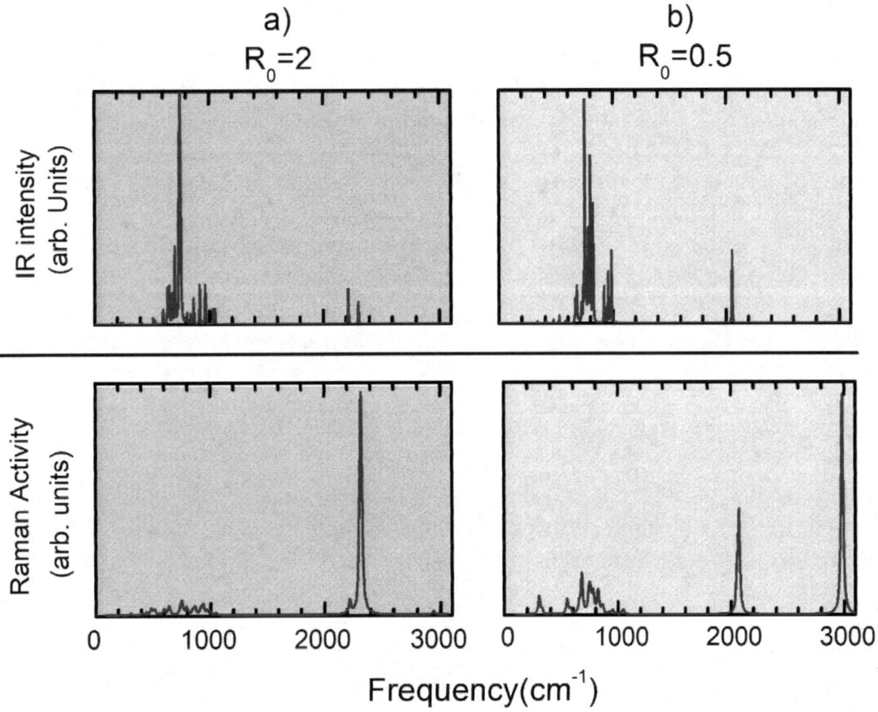

Figure 3. IR (top panel) and Raman (bottom panel) spectrum of porous Silicon Carbide with porosity P=12.5%, and a) R_0=2 and b) R_0=0.5.

4. Conclusions

In this work the phonon band structure, IR and Raman spectroscopies of porous silicon carbide with Si or C rich surfaces, were studied using the first principles density functional perturbation theory. Results show that there are noticeable differences between the Si and C-rich surfaces, as the C-Rich surface has more packed bands with two band gaps at low frequencies, while the Si-Rich one has something more resembling to a continuum of bands at the same frequency intervals. Noticeable differences appear between both systems at the high frequency bond stretching regions, where the differences between Si-Rich and C-rich systems are notable .These divergences are in turn reflected in the Raman and IR spectrum where the IR allows better characterization of the low frequency modes while the Raman is better for high frequency modes. The disparities between the two systems in Raman and IR spectrum, and the absence of negative modes in their vibrational modes could allow for a better experimental characterization, which in turn will be useful for the use of pSiC in devices such as supercapacitors.

5. References

[1] Simon P, Gogotsi Y and Dunn B 2014 *Science* **343** 1210-1
[2] Kim M, Oh I and Kim J 2016 *Electrochimica Acta* **196** 357-68
[3] Kim M, Oh I and Kim J 2016 *Chemical Engineering Journal* **289** 170-9
[4] Nangir M and Massoudi A 2020 *Materials Today: Proceedings*
[5] Sarno M, Galvagno S, Scudieri C, Iovane P, Portofino S, Borriello C and Cirillo C 2019 *Journal of Physics and Chemistry of Solids* **131** 213-22
[6] Mourya S, Kumar A, Jaiswal J, Malik G, Kumar B and Chandra R 2019 *Sensors and Actuators B: Chemical* **283** 373-83
[7] Calvino M, Trejo A, Crisóstomo M C, Iturrios M I, Carvajal E and Cruz-Irisson M 2016 *Theoretical Chemistry Accounts* **135** 104
[8] Calvino M, Trejo A, Iturrios M I, Crisóstomo M C, Carvajal E and Cruz-Irisson M 2014 *Journal of Nanomaterials* **2014** 1-7
[9] González I, Trejo A, Calvino M, Miranda A, Salazar F, Carvajal E and Cruz-Irisson M 2018 *Physica B: Condensed Matter* **550** 420-7
[10] Refson K, Tulip P R and Clark S J 2006 *Physical Review B* **73** 155114
[11] Perdew J P, Burke K and Ernzerhof M 1996 *Physical Review Letters* **77** 3865-8
[12] Hamann D R, Schlüter M and Chiang C 1979 *Physical Review Letters* **43** 1494-7
[13] Monkhorst H J and Pack J D 1976 *Physical Review B* **13** 5188-92
[14] de Santiago F, Santana J E, Miranda Á, Trejo A, Vázquez-Medina R, Pérez L A and Cruz-Irisson M 2019 *Applied Surface Science* **475** 278-84
[15] González I, Sosa A N, Trejo A, Calvino M, Miranda A and Cruz-Irisson M 2018 *Dalton Transactions* **47** 7505-14
[16] Shin W, Seo W, Takai O and Koumoto K 1998 *Journal of Electronic Materials* **27** 304-7
[17] Trejo A, Miranda A, Toscano-Medina L K, Vázquez-Medina R and Cruz-Irisson M 2016 *Microelectronic Engineering* **159** 215-20

Acknowledgments
This work was supported by projects from SIP-IPN, computations were carried out at Laboratorio Nacional de Supercómputo del Sureste de México of BUAP

IOP Conf. Series: Materials Science and Engineering **840** (2020) 012010 doi:10.1088/1757-899X/840/1/012010

Visible and Infrared Photoluminescence in Hexagonal Silicon Carbide by Direct Femtosecond Laser Writing

Stefania Castelletto [1,2,*], Brett C Johnson [3], Alberto Boretti [4,*]

[1]School of Engineering, RMIT University, Bundoora, Victoria, Australia
[2]Swinburne Nanotechnology facility, Centre for Micro-Photonics, Swinburne University of Technology, Hawthorn, Victoria, Australia
[3]School of Physics, The University of Melbourne, Victoria, Australia
[4]Department of Mechanical Engineering, College of Engineering, Prince Mohammad Bin Fahd University, Al Khobar, Saudi Arabia

*Email: a.a.boretti@gmail.com;stefania.castelletto@rmit.edu.au

Abstract. Optically active color centers in silicon carbide have attracted considerable attention in the past few years as candidates for quantum technologies such as single-photon sources, nanomagnetic resonance imaging, and spintronic devices. Control over defect position and their placement at the desired location within a chip, necessary to integrate them within optical and electronic devices, is still a challenge. Recently, laser writing emerged as a new tool to generate vacancies in crystals as a starting point for the formation of color centers. In this work, a laser writing method has been used to produce color centers in 4H and 6H bulk silicon carbide by using a femtosecond laser. An array of color centers was fabricated by different pulse laser energies in sites of square grids at varying depths (from the surface to 10 μm below the surface). We optically characterized the fabricated color centers using confocal imaging, photoluminescence, and Raman spectroscopy. We show that the technique can produce specifically vacancy color centers with a relevant emission in the visible (peak around 700 nm) and near-infrared (peak at 900 nm) with the latter identified as the silicon-vacancy. This method can be adapted to engineer color centers in silicon carbide at different depths in the material, for the above-mentioned applications, in addition to the fabrication of light-emitting diodes.

1. Introduction

Laser writing is receiving more and more attention to the creation of the point defects used in photonics and quantum devices. Laser writing may produce point defects in transparent semiconductors or dielectric materials, following different physical nonlinear effects that are not fully understood for each specific material. Silicon carbide (SiC) is a relevant transparent wide bandgap semiconductor, due to its large bandgap, heteroepitaxial growth on Silicon, over 200 crystal polytypes, high hardness, and high voltage breakdown, wafer-scale, and CMOS compatibility; qualities that make it a key material for sensors, electronic devices and more recently photonics [1]. Silicon carbide has also been recently studied as a quantum material and for its application in quantum spintronics [2] and quantum technologies [3].

The quantum properties in SiC are associated with fluorescent color centers [4] that can produce single-photon sources and spin qubits. Other applications are sensitive magnetometry [5] and thermometry [6], with the prospects of color centers operating in the near-infrared region for deep tissue imaging by its nanostructuring and on-demand placement of color centers [7]. Most of the optically and

IOP Conf. Series: Materials Science and Engineering **840** (2020) 012010 doi:10.1088/1757-899X/840/1/012010

spin active color centers in SiC are vacancy-related emitters, such as the simplest silicon-vacancy (V_{Si}) [8], [9]. Other vacancy-related emission bands are due to complex vacancy structures such as the carbon antisite vacancy pair [10] and divacancies [11]. These color centers can be generated by electron irradiation [10], [12], ions beams [13, 14], neutrons [8], [11], proton irradiation [15] and focused ion beams [16] with subsequent annealing at specific temperatures to differentiate the diffusion of vacancies. While focused ions beam and proton irradiation allow creating an array of emitters without the use of a mask, other methods either need a mask or do not have correct localization of the emitters.

Recently another successful method to create color centers in optical materials has been based on the use of femtosecond laser writing, whereas single color centers in diamond [17], boron nitride [18], [19] and SiC have been demonstrated [20], [21] from ensemble to the single defect level. The advantages of using femtosecond lasers for micro and nano machining of bulk optically transparent materials, rely on the possibility to fabricate geometrically complex structures in three dimensions and compound substrates of different materials [22].

In this paper, we summarize the optical properties of direct laser written 4H and 6H SiC, where luminescent centers associated with vacancies in the visible and near-infrared are formed while keeping the material damage minimal as shown by Raman spectroscopy maps.

2. Methods

The methods and material fabrication are fully described in [20]. A 1,030 nm wavelength, 230 fs duration and a repetition rate of 200 kHz laser were used to create a vacancy in 4H and 6H SiC 60 µm epilayers of 4H and 6H SiC unintentionally doped. Part of the measurements collected but not included in that paper are here presented, together with a reanalysis of the past measurements for the physical process that is involved in the material vacancy creation.

3. Results

Materials fabrication and methods were previously described [20]. The laser writing pattern is of 10x10 array of dots with 5 µm pitch, with each line corresponding to a tuned laser energy per single pulse, from 230 nJ to 0.45 nJ. Laser writing was achieved with a single pulse directly on the surface of the samples, 5 µm and 10 µm depth from the surface. We characterized the laser written samples after a surface cleaning using Piranha solution, sulfuric acid, and HF etch to remove surface color centers and other organic defects, which could limit the detection of the deep generated defects.

For the characterization of the generated emitters, we used Raman spectroscopy, fluorescence microscopy, and lifetime measurements. Confocal images of a laser written area in 4H-SiC are shown in Figure (a-b-c) using 532 nm and 633 nm excitation and in both cases a long-pass filter at 780 nm.

In Figure 1 we show the confocal image of the laser written emitters excited with 532 nm and 633 nm in 4H SiC at different depths. We saw a broad red emission centered at 700 nm for 230 nJ to 28.8 nJ energy laser written dots on the area written close to the surface, while emission centered at 900 nm was seen only up to laser written dots of 57.6 nJ /pulse. We observe a photoluminescence (PL) intensity scaling with laser energy, showing that the emission is due to laser irradiation. Using the 633 nm excitation the emission at 900 nm is more efficiently excited.

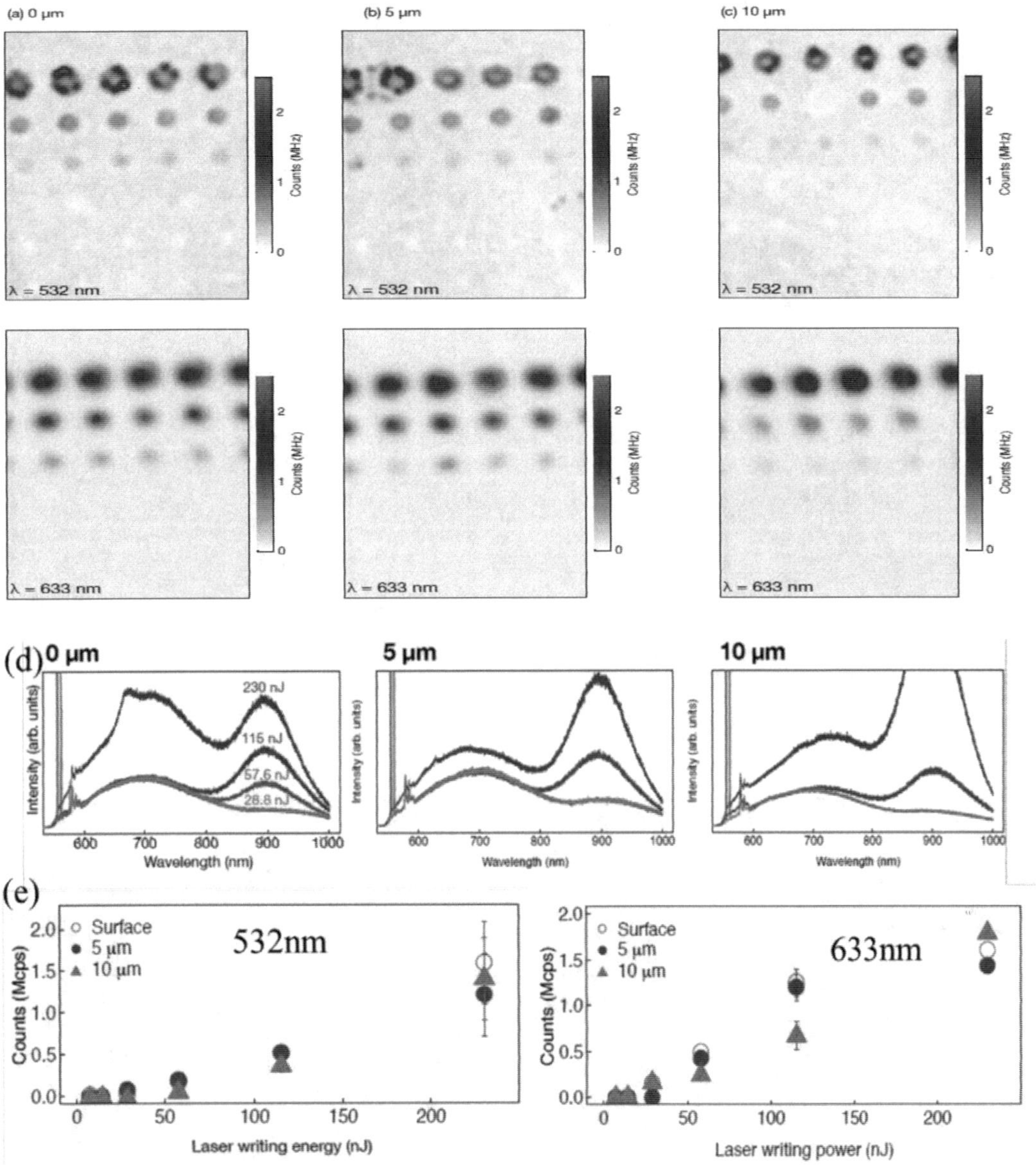

Figure 1. 532 nm and 633 nm excitations confocal images of a laser written 4H SiC surface using a 780 nm long-pass filter for fabrication at the surface (a), 5μm (b) and 10μm (c). Dots correspond to photoluminescence created using a laser pulse at 28.8, 57.5, 115, and 230 nJ /pulse. (d) PL from laser written dots at the surface, 5μm and 10μm below surface showing an emerging emission attributed to the V_{Si} centered at 900 nm together with a broad PL centered at 700 nm. Raman peaks appear under 600 nm. (e) PL count rate of the spots extracted from confocal images. An increase of count rate is seen as the laser writing energy increases for the three laser writing depths: at the surface and 5 and 10 μm below the surface. Fig. 3 of [20] only reports a.1, d.1 and e.1, part of the 532 nm results, and not the 633 nm results. Figures a.1, d.1 and e.1 adapted with permission from [20] © The Optical Society.

In Figure 2 we show the Raman modes of the pristine material and laser written one and the maps of the Raman intensity (c), the Raman shift in peak position (d) and broadening (e) for the mode TO(E_2) at different depths and the mode LO (A_1). The reduction of the Raman line intensity is due likely to scattering from the surface where ablation occurred, while a Raman frequency shift is seen due to change in carrier concentration due to laser multiphoton ionization or avalanche ionization. The LO (A1) peak showed greater shifts and broadening as it is more sensitive to changes in the carrier concentration as evident from the maps in Figure 2 (g).

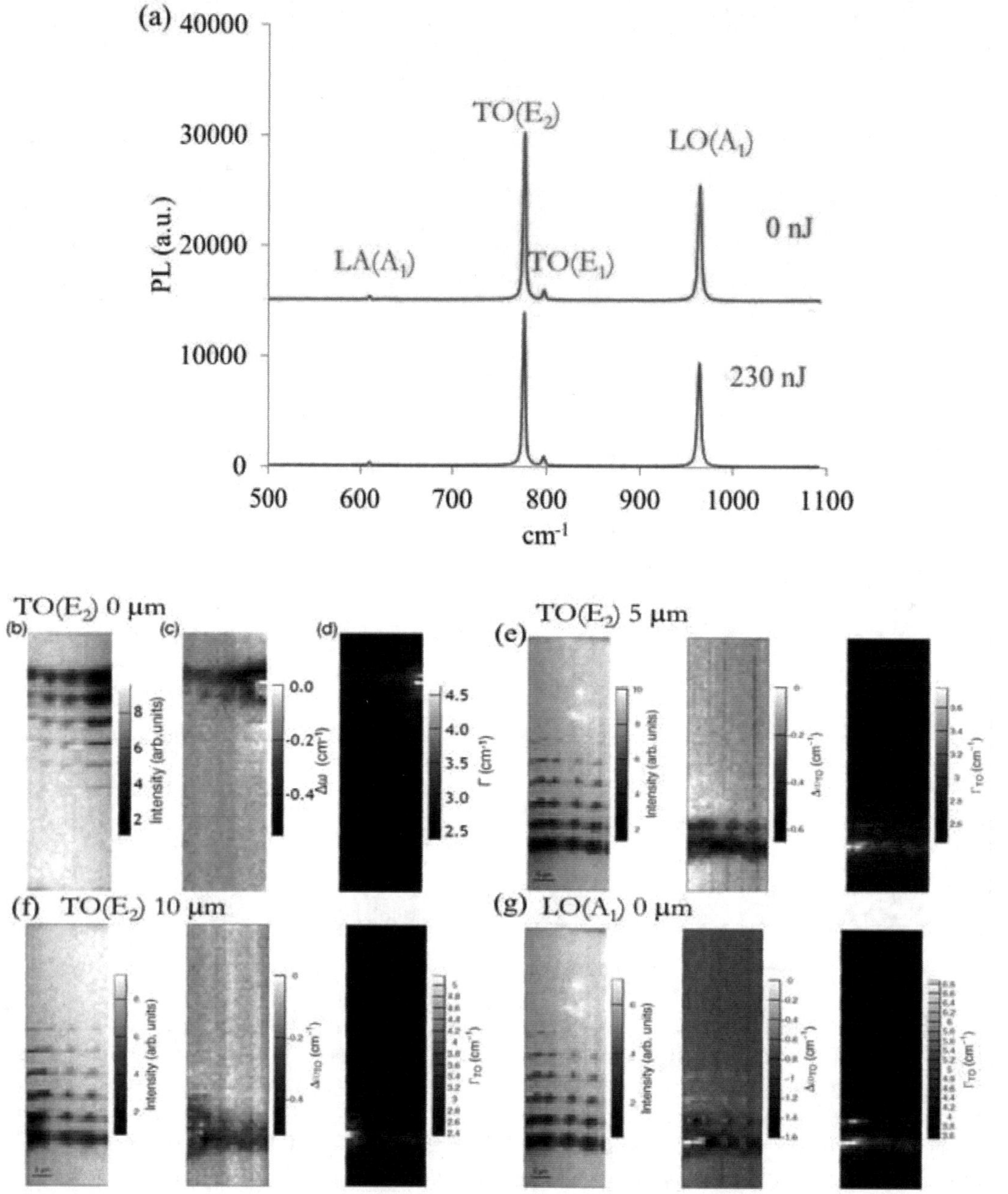

Figure 2. (a) Raman spectra (top) of the unmodified 4H-SiC surface and from a 230 nJ laser-written spot (bottom). The spectra are offset vertically for clarity. The vibrational modes are labeled. (b) to (f) Raman maps of the laser written areas obtained with 532 nm excitation of the surface TO(E_2) peak fit intensity (b), a shift in frequency of the peak position (c) and its broadening (d), all determined by fitting the TO(E_2) emission. (e) Same Raman maps for the TO (E_2) at 5 µm and (f) at 10 µm below the surface. (Fig. 2 of [20] reports similar results based on different measurement campaign).

In Figure 3 the PL decays were measured from the emission of the fabricated areas on the surface and at different depths by using a pulsed laser at 730 nm excitation with a 40 MHz repetition rate and collecting with a long pass filter at 850 nm. For comparison, the PL decay of a bulk 4H SiC electron irradiated at 10^{18} e/cm^2 with V_{Si} ensemble emission is also shown as a reference. For the ensemble V_{Si} a single exponential decay is seen and a measured lifetime of 6.3 ns. For the fabricated areas, the PL decays were fitted using a double exponential curve, with the longest decay rates ranging from 2.3 ns (for surface emitters) to up to 11 ns (for 10 μm below surface emitters). We saw a longer lifetime after annealing at 300 °C in Argon of the same fabricated dots. Longest lifetime was associated to lower formation energy (57 nJ) at the depth of 10 μm, while shorter lifetime was seen for 230 nJ formation energy and on the surface. This is consistent with the expected better optical properties of emitters generated deeper in the material. The in general shorter lifetime compared to electron irradiated can be attributed to some non-radiative quenching due to the high concentration of defects created, residual damage or the material doping. Further study of the interaction of the laser energy with the pristine material properties should be performed.

Comparable results were seen in the 6H SiC, however, in this material fabrication was carried out only close to the surface. In Figure 4, we show the confocal image of the written area in the 6H SiC using 532 nm excitation compared to an SEM image of the same area. The spectral analysis shows the formation of emission at 900 nm however in this case the brightness was lower compared to the 4H material. The lifetime of the same 920 nm emitters obtained in 6H SiC with the same procedure, confirm comparable results shown for 4H SiC.

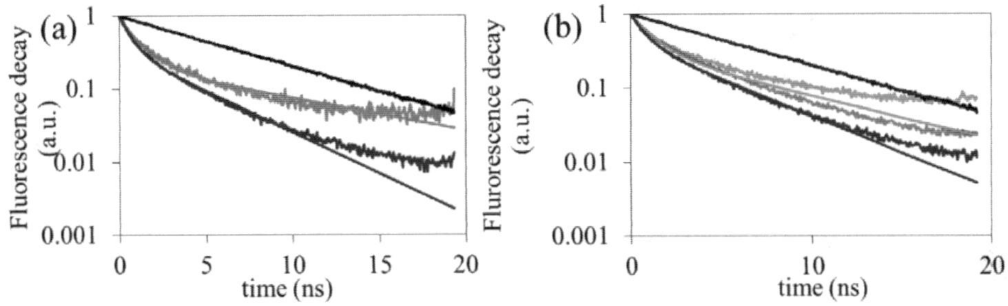

Figure 3. The lifetime of the laser written 4H SiC dots at 57 nJ at 10 μm (green), 115 nJ at 10 μm (red) and 230 nJ on the surface (blue), corresponding to a two-exponential decay fit, before annealing (3.9ns for 230 nJ, 4.5 ns and 115 nJ) (a) and after annealing (4.4 ns for 230 nJ, 5.2 ns for 115 nJ, 6.9ns for 57 nJ) (b). The dots that are written with 57 nJ before annealing were not observed. The black line corresponds to the lifetime of an ensemble of electron emitters corresponding to 6.3ns.

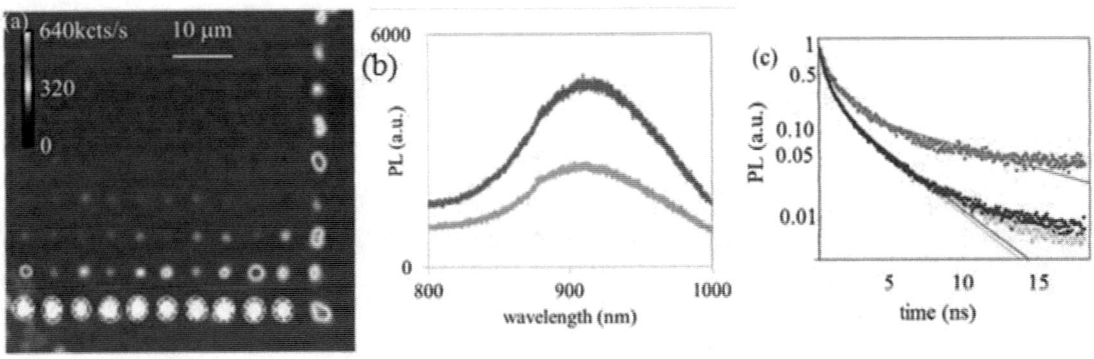

Figure 4. (a) Confocal image using 532 nm excitation and a 650 nm long-pass filter of the laser written dots area of the 6H-SiC. From the bottom brightest emission corresponds to dots fabricated with 230 nJ

/pulse, 115 nJ /pulse, 57 nJ /pulse, and 28.8 nJ /pulse. (b) PL emission excited at 532 nm for the dots written with 230 nJ /pulse (top) and 115 nJ /pulse (bottom). (c) The lifetime of the 6H laser written dots on the surface at 57 nJ (red), 115 nJ (blue) and 230 nJ (green), corresponding to a two-exponential decay fit with longer exponential of 8.7, 3.3, 3.1 ns, respectively. Annealing was not performed in this sample.

4. Summary and conclusions

In this work, we show the successful generation of color centers in 4H and 6H SiC using a sub-picosecond pulsed laser at 1030 nm, attributed to the silicon-vacancy for emission around 900 nm. The pulse energy formation observed here is quite high. It is 57 nJ/pulse, corresponding to approximately the generation of ~10 color centers based on the saturation count rate measured. The emission with a peak at 700 nm is not identified, however, it is a damage related emission observed even at lower energy 28 nJ /pulse. The laser energy threshold, in this case, is higher than what is observed in [17] which is about 12.6 nJ for ensemble generation, and the depth in this work in at the most 10 μm. No single-photon emission was observed in these samples due to large damage and or large background. These results are related to the mechanism of laser writing of vacancies in transparent wide-bandgap materials. For a transparent wide bandgap material illuminated by an intense femtosecond laser, many excited free electrons are generated that may produce intrinsic defects, resulting in photoinduced damage. Nonlinear photon absorption mechanisms are responsible for the generation of free electrons according to two models, known as tunneling ionization (Zener breakdown) and multiphoton ionization (MPI) [23]. The dominant mechanism for free-electron generation is typically delineated by the Keldysh parameter [23]. For our experiment, the laser intensity for MPI has to be less than 1.8 10^{17} W/m^2, that considering a pulse of 230 fs and a beam waist at the focus of the 0.9 NA objective of 364 nm, corresponds to 17nJ/pulse. As we did not observe emission from less than 28.8 nJ/pulse, we are not in the MPI regime rather in the tunneling ionization regime and avalanche photoionization can occur. Further, the exact concentration of electrons in the conduction band in the starting material makes the optical breakdown threshold and damage non-deterministic [23]. Work is currently in progress to optimize the substrate and laser fabrication methods.

5. References

[1] Xing P, Ma D, Ooi K J A, Choi J W, Agarwal A M and Tan D 2019 CMOS-Compatible PECVD Silicon Carbide Platform for Linear and Nonlinear Optics *ACS Photonics* 6 1162-7
[2] Awschalom D D, Hanson R, Wrachtrup J and Zhou B B 2018 Quantum technologies with optically interfaced solid-state spins *Nature Photonics* 12 516-27
[3] Atatüre M, Englund D, Vamivakas N, Lee S-Y and Wrachtrup J 2018 Material platforms for spin-based photonic quantum technologies *Nature Reviews Materials* 3 38-51
[4] Lohrmann A, Johnson B C, McCallum J C and Castelletto S 2017 A review on single photon sources in silicon carbide *Reports on Progress in Physics* 80 034502
[5] Simin D, Soltamov V A, Poshakinskiy A V, Anisimov A N, Babunts R A, Tolmachev D O, Mokhov E N, Trupke M, Tarasenko S A, Sperlich A, Baranov P G, Dyakonov V and Astakhov G V 2016 All-Optical dc Nanotesla Magnetometry Using Silicon Vacancy Fine Structure in Isotopically Purified Silicon Carbide *Physical Review X* 6 031014
[6] Kraus H, Soltamov V A, Fuchs F, Simin D, Sperlich A, Baranov P G, Astakhov G V and Dyakonov V 2014 Magnetic field and temperature sensing with atomic-scale spin defects in silicon carbide *Scientific reports* 4 5303-
[7] Castelletto S, Al Atem A S, Inam F A, Bardeleben H J v, Hameau S, Almutairi A F, Guillot G, Sato S-i, Boretti A and Bluet J M 2019 Deterministic placement of ultra-bright near-infrared color centers in arrays of silicon carbide micropillars *Beilstein Journal of Nanotechnology* 10 2383-95
[8] Fuchs F, Stender B, Trupke M, Simin D, Pflaum J, Dyakonov V and Astakhov G V 2015 Engineering near-infrared single-photon emitters with optically active spins in ultrapure silicon carbide *Nature Communications* 6 7578

[9] Widmann M, Lee S Y, Rendler T, Son N T, Fedder H, Paik S, Yang L P, Zhao N, Yang S, Booker I, Denisenko A, Jamali M, Momenzadeh S A, Gerhardt I, Ohshima T, Gali A, Janzen E and Wrachtrup J 2015 Coherent control of single spins in silicon carbide at room temperature *Nat. Mater.* 14 164

[10] Castelletto S, Johnson B C, Ivady V, Stavrias N, Umeda T, Gali A and Ohshima T A 2014 Silicon carbide room-temperature single-photon source *Nat. Mater.* 13 151

[11] Christle D J, Falk A L, Andrich P, Klimov P V, Hassan J U, Son N T, Janzén E, Ohshima T and Awschalom D D 2015 Isolated electron spin in silicon carbide with millisecond coherence times *Nat. Mater.* 14 160

[12] Christle D J, Klimov P V, de las Casas C F, Szász K, Ivády V, Jokubavicius V, Ul Hassan J, Syväjärvi M, Koehl W F, Ohshima T, Son N T, Janzén E, Gali Á and Awschalom D D 2017 Isolated Spin Qubits in SiC with a High-Fidelity Infrared Spin-to-Photon Interface *Physical Review X* 7 021046

[13] Falk A L, Buckley B B, Calusine G, Koehl W F, Dobrovitski V V, Politi A, Zorman C A, Feng P X L and Awschalom D D 2013 Polytype control of spin qubits in silicon carbide *Nature communications* 4 1819-

[14] Rühl M, Ott C, Götzinger S, Krieger M and Weber H B 2018 Controlled generation of intrinsic near-infrared color centers in 4H-SiC via proton irradiation and annealing *Applied Physics Letters* 113 122102

[15] Kraus H, Simin D, Kasper C, Suda Y, Kawabata S, Kada W, Honda T, Hijikata Y, Ohshima T, Dyakonov V and Astakhov G V 2017 Three-Dimensional Proton Beam Writing of Optically Active Coherent Vacancy Spins in Silicon Carbide *Nano Letters* 17 2865-70

[16] Wang J, Zhang X, Zhou Y, Li K, Wang Z, Peddibhotla P, Liu F, Bauerdick S, Rudzinski A, Liu Z and Gao W 2017 Scalable fabrication of single silicon vacancy defect arrays in silicon carbide using focused ion beam *ACS Photonics* 4 1054

[17] Chen Y-C, Griffiths B, Weng L, Nicley S S, Ishmael S N, Lekhai Y, Johnson S, Stephen C J, Green B L, Morley G W, Newton M E, Booth M J, Salter P S and Smith J M 2019 Laser writing of individual nitrogen-vacancy defects in diamond with near-unity yield *Optica* 6 662-7

[18] Songyan H, Muhammad Danang B, Saleem U, Maurice Ange A, Roland Yingjie T, Philippe C, Beng Kang T, Hong W and Edwin Hang Tong T 2018 Localized emission from laser-irradiated defects in 2D hexagonal boron nitride *2D Materials* 5 015010

[19] Buividas R, Aharonovich I, Seniutinas G, Wang X W, Rapp L, Rode A V, Taniguchi T and Juodkazis S 2015 Photoluminescence from voids created by femtosecond-laser pulses inside cubic-BN *Opt. Lett.* 40 5711-3

[20] S. Castelletto, A. F. M. Almutairi,, K. Kumagai, T. Katkus, Y. Hayasaki, B. C. Johnson, and S. Juodkazis 2018 Photoluminescence in hexagonal silicon carbide by direct femtosecond laser writing *Opt. Lett* 43(24) 6077-6080. doi.org/10.1364/OL.43.006077.

[21] Chen Y-C, Salter P S, Niethammer M, Widmann M, Kaiser F, Nagy R, Morioka N, Babin C, Erlekampf J, Berwian P, Booth M J and Wrachtrup J 2019 Laser Writing of Scalable Single Color Centers in Silicon Carbide *Nano Letters* 19 2377-83

[22] Malinauskas M, Žukauskas A, Hasegawa S, Hayasaki Y, Mizeikis V, Buividas R and Juodkazis S 2016 Ultrafast laser processing of materials: from science to industry *Light: Science & Applications* 5 e16133-e

[23] Schaffer C. B., Brodeur A., Mazur E. 2001 Laser-induced breakdown and damage in bulk transparent materials induced by tightly focused femtosecond laser pulses. *Meas. Sci. Technol.* 12, 1784.

Acknowledgments
We acknowledge the Swinburne Nanotechnology facility for fabricating the laser written samples and the RMIT node of the Australian Research Council (ARC) Center for Nano Biophotonics team for

providing lifetime measurements facility. BCJ acknowledges the University of Melbourne ARC for quantum computing.

IOP Conf. Series: Materials Science and Engineering **840** (2020) 012011 doi:10.1088/1757-899X/840/1/012011

Mechanical and Thermal Behaviour of Epoxy/Hexagonal Boron Nitride/Short Sisal Fiber Hybrid Composites

Alok Agrawal[1], Saurabh Chandrakar[2], Abhishek Sharma[3]

[1]Mechanical Engineering Department, Sagar Institute of Research and Technology Excellence, Bhopal-462041, Madhya Pradesh, India
[2]Department of Mechanical Engineering, National Institute of Technology Karnataka, Surathkal-575025, India
[3]Mechanical Engineering Department, Manipal University Jaipur, Jaipur, Rajasthan, India

Email: alokag03@gmail.com

Abstract. Hybrid composite i.e. surface modified hexagonal boron nitride (hBN) and short sisal fiber reinforced in epoxy matrix is fabricated using hand lay-up method. The effect of surface modified hBN filler and sisal fiber content on mechanical and thermal properties of epoxy based hybrid composites were investigated in this paper. The main aim of the investigation is to develop a material which can found its application in microelectronic components. As per the requirement of microelectronic industry, the material should possess high thermal conductivity. Hence, thermal conductivity of epoxy increases with increase in hexagonal boron nitride content. Inspite of insulative nature of sisal fiber, the study shows that its inclusion in combination with hBN enhances the thermal conductivity if the content of both the fillers were properly selected. Other thermal property like coefficient of thermal expansion and glass transition temperature appreciably improves when combination of fillers were added in epoxy matrix. Mechanical properties under study i.e. tensile strength and compressive strength also enhances when combination of sisal fiber and hBN were incorporated as compared to when single filler hBN were used. Hence, usage of hybrid filler as reinforcement in epoxy improve overall mechanical and thermal property of the developed material.

1. Introduction

With rapid development in modern microelectronic industries, there is a continuous increase in power density of the electronic devices. The power density increases mainly because of everlasting trend towards miniaturization of such devices. Miniaturization results in overheating of the device and hence will cause thermal problem if the generated heat will not dissipated effectively and efficiently. This further cause many other problems in the performance of the device like power leakage, deterioration of performance and even device failure [1]. Apart from reasonably high thermal conductivity for heat dissipation, other requirement of material to be used in electronic devices are low dielectric constant for proper signal distribution with high speed or low delay time and to diminish probability of short circuiting, low coefficient of thermal expansion to avoid thermal fatigue, high thermal stability to withstand sudden rise of temperature, high glass transition temperature to maintain its crystal structure at elevated temperature and reasonably good mechanical properties to protect them from external damages or impact failure. All the mentioned properties to be achieved in monolithic material are difficult. With this, demand of reliable and high performance packaging material is increasing. Thus,

IOP Conf. Series: Materials Science and Engineering **840** (2020) 012011 doi:10.1088/1757-899X/840/1/012011

proper selection of material is necessary to enhance device efficiency and protect them from premature failure [2].

This leads to the demand of developing composite material. Polymers have excellent dielectric properties with added advantage of ease of process ability at low cost. But because of low thermal conductivity of polymers, their applications in electronic components are limited [3]. Ceramic fillers with good thermal conductivity and low dielectric constant are of interest among the researchers to be used as filler material in polymeric matrix. The various ceramic fillers used in different polymeric matrix for similar applications are silicon carbide [4], aluminium oxide [5], aluminium nitride [6], boron nitride [7], montmorillonite [8] etc. In all the above study, filler content govern the thermal conductivity of the developed material. However, increasing the filler content to higher value results in brittle material with low melt flow index, increased porosity, increase in water absorption rate and have adverse effect on dielectric properties [9]. Thus, to design and develop an electrically insulated and high thermal conductive composite without comprising its mechanical properties is important.

In view of this, in present investigation, hybrid filler were used in epoxy matrix to develop such composites. In the process of developing such composites, it has been kept in mind to use material which is environmental friendly. Synthetic and natural fibers reinforcement in polymeric composites is reported to increase the mechanical behavior by various researchers [10-12]. They are electrically insulated as well. Hence, a natural fiber i.e. sisal fiber is utilized as reinforcement in epoxy matrix together with ceramic filler i.e. hexagonal boron nitride (hBN). Sisal fiber is used mainly to provide ductile behavior to material which would have been brittle in nature if only hBN were used in epoxy matrix as seen in previous work. Also, sisal fiber is highly insulative which controls the dielectric property of the material. The main focus of this work is to investigate the improvement in mechanical and thermal behavior provided by the addition of natural fiber to the hBN/epoxy composites.

2. Experimental details

Matrix material used in present investigation is Lapox L12 with its corresponding hardener K6. The matrix material used is supplied by ATUL India Ltd., Gujarat, India. Hexagonal boron nitride of size 5 microns is used as primary filler material. It is procured from Souvenier Chemicals, Mumbai. It has density of 2.3 g/cc. Sisal fiber in short form i.e. around 4 mm length is used as second filler. The sisal fiber possesses low density of 1.41 g/cc and good tensile strength of approx. 350-370 MPa [13].

Hand lay-up technique is implemented for preparation of epoxy based hybrid composites. The resin and hardener used in present investigation is combined in 10:1 ratio by weight. Surface modified hexagonal boron nitride and short sisal fiber were added to the epoxy-hardener solution. The combination is than mixed manually and slowly. Silane coupling agent of 2wt. % concentration in ethanol is used for surface treatment of hBN [9] and 2 mole aqueous solution of sodium hydroxide is used for surface treatment of sisal fiber [14]. Prepared mixture is than poured slowly in to the mould. Before pouring the mixture in to the mould, coating of silicon spray is mandatory for easy removal of composite after it gets cured. Cast take duration of 8 hours before it gets completely cured and ready for removal from the mould. The list of fabricated composite is presented in table 1.

Tensile strength of the samples is measured as per ASTM D638 and compressive strength is measured as per ASTM D695 standard. Both the tests were performed in Tinius Olsen universal testing machine. Thermal conductivity is measured as per ASTM E-1530 with the help of Unitherm™ Model 2022. Perkin-Elmer thermal mechanical analyzer (TMA-7) is used to measure Coefficient of thermal expansion and glass transition temperature as per ASTM D 618 and ASTM E 831 respectively.

3. Results and discussions

3.1. Tensile strength

The variation in tensile strength on incorporation of hexagonal boron nitride and sisal fiber in epoxy matrix is presented in Figure 1. It can be seen from the figure that when 10 wt. % hBN is added in epoxy matrix, the tensile strength increased from 40.5 MPa for neat epoxy to 45.2 MPa.

Table 1. List of fabricated specimens.

Set	Composition	Set	Composition	Set	Composition	Set	Composition
Set A1	Epoxy + 10 wt % hBN	**Set B1**	Epoxy + 20 wt % hBN	**Set C1**	Epoxy + 30 wt % hBN	**Set D1**	Epoxy + 40 wt % hBN
Set A2	Epoxy + 10 wt % hBN + 1 wt % SF	**Set B2**	Epoxy + 20 wt % hBN + 1 wt % SF	**Set C2**	Epoxy + 30 wt % hBN + 1 wt % SF	**Set D2**	Epoxy + 40 wt % hBN + 1 wt % SF
Set A3	Epoxy + 10 wt % hBN+ 2 wt % SF	**Set B3**	Epoxy + 20 wt % hBN+ 2 wt % SF	**Set C3**	Epoxy + 30 wt % hBN+ 2 wt % SF	**Set D3**	Epoxy + 40 wt % hBN+ 2 wt % SF
Set A4	Epoxy + 10 wt % hBN+ 3 wt % SF	**Set B4**	Epoxy + 20 wt % hBN+ 3 wt % SF	**Set C4**	Epoxy + 30 wt % hBN+ 3 wt % SF	**Set D4**	Epoxy + 40 wt % hBN+ 3 wt % SF
Set A5	Epoxy + 10 wt % hBN+ 4 wt % SF	**Set B5**	Epoxy + 20 wt % hBN+ 4 wt % SF	**Set C5**	Epoxy + 30 wt % hBN+ 4 wt % SF	**Set D5**	Epoxy + 40 wt % hBN+ 4 wt % SF
Set A6	Epoxy + 10 wt % hBN+ 5 wt % SF	**Set B6**	Epoxy + 20 wt % hBN+ 5 wt % SF	**Set C6**	Epoxy + 30 wt % hBN+ 5 wt % SF	**Set D6**	Epoxy + 40 wt % hBN+ 5 wt % SF

It has been observed in earlier investigation also that a relatively small quantity of filler can successfully remedy the defects from epoxy self-curing. However, when hBN content increases further, decreasing trend in tensile strength is observed. For maximum hBN loading of 40 wt. %, the tensile strength of composites decreased to 30.8 MPa.

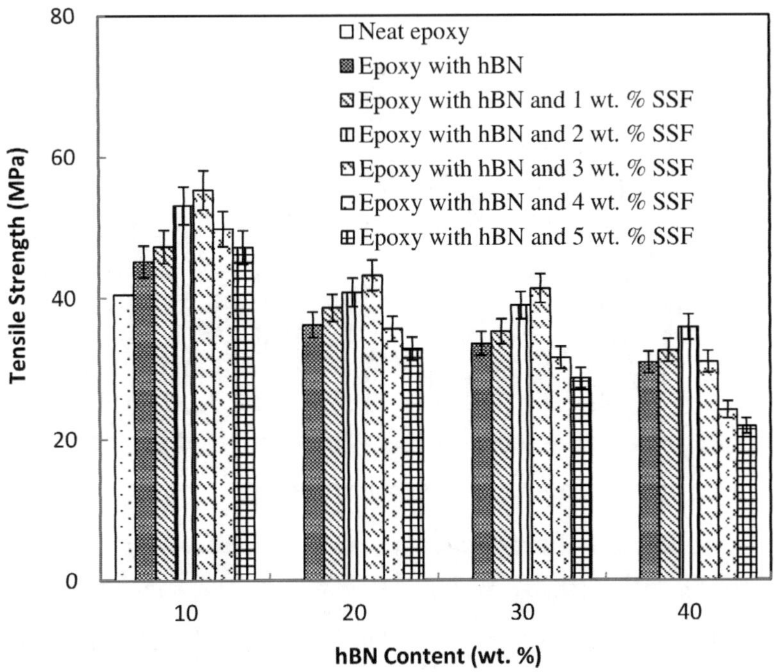

Figure 1. Tensile strength of fabricated samples.

The reduction in tensile strength with high filler loading is due to fact that excessive fillers served as local stress-concentration points in the resin matrix causing in the weakening in tensile strength.

Again, when combination of sisal fiber and hBN is incorporated, it successfully enhances the tensile strength of epoxy matrix but only when the content of sisal fiber is limited. It is observed that when sisal fiber were added in combination with hBN in epoxy, tensile strength of the hybrid composite increases upto 3 wt. % of sisal fiber. When content of fiber increases further, decreasing trend in the tensile strength is observed. Incorporation of 3 wt. % of fiber in combination with 10 wt. % of hBN shows maximum tensile strength of 55.3 MPa. Further increase of sisal fiber decreases the tensile strength which reduces to 47.2 MPa for 5 wt. % sisal fiber.

3.2. Compressive strength

The variation in compressive strength of epoxy composites filled with different content of hexagonal boron nitride and sisal fiber is shown in figure 2. With hBN as single filler is incorporated in epoxy matrix, compressive strength increases from 85 MPa for neat epoxy to 108.7 MPa for maximum loading of 40 wt. % hBN in epoxy which is around 27.9 % increment in compressive strength. This increment increases further when sisal fiber were added in combination with hBN in epoxy matrix. In this case, when 5 wt. % of fiber were added in combination with 40 wt. % hBN in epoxy matrix, compressive strength increases from 108.7 MPa to 118.1 MPa which is an increment of 38.9 % over neat epoxy. Unlike tensile strength, compressive strength always show increasing trend with increase in filler content. The increasing behaviour in compressive strength is attributed to the favorable deformation processes which are facilitated due to the presence of hBN particles and short sisal fiber in the epoxy matrix.

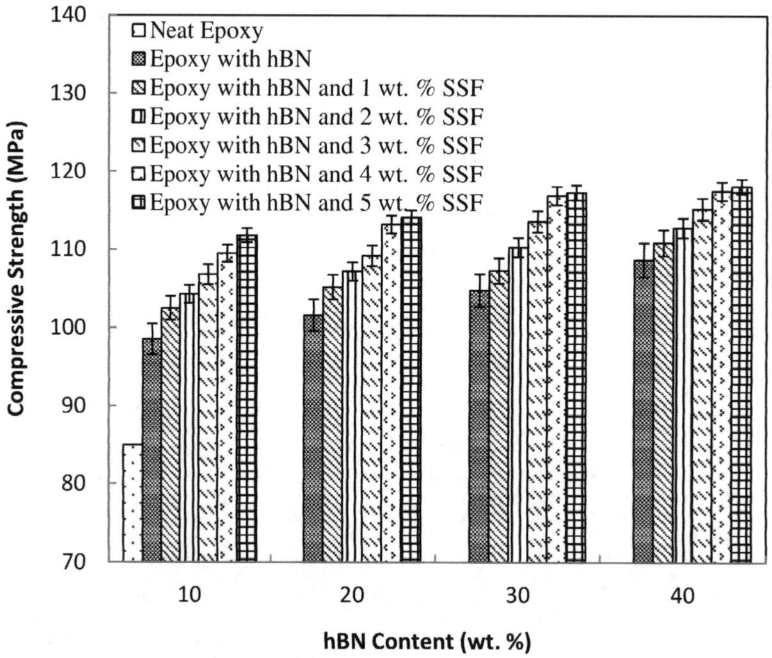

Figure 2. Compressive strength of fabricated samples.

3.3. Thermal conductivity

The variation in thermal conductivity of epoxy composites filled with different content of hexagonal boron nitride and sisal fiber is shown in figure 3. When hBN content was 40 wt. % in epoxy, the thermal conductivity was increased from 0.211 W/m-K for neat epoxy to 1.52 W/m-K. Further when sisal fiber is added in epoxy/hBN combination, the thermal conductivity decreases with increase in sisal fiber upto 2 wt. %. Later, the highest value of thermal conductivity is obtained with a combination of 30 wt. % hBN and 3 wt. % of sisal fiber. The value reported is 1.88 W/m-K. Also improved value of thermal conductivity is obtained for 40 wt. % hBN and 3 wt. % of sisal fiber which is having a value of 1.55 W/m-K. Apart from the above combination, thermal conductivity decreases with sisal fiber content. This may be attributed to the fact that hBN was well distributed and ordered on the sisal fiber surfaces for a particular combination. This arrangement helped the hBN powder interconnection, resulting in a significant improvement in thermal conductivity [15].

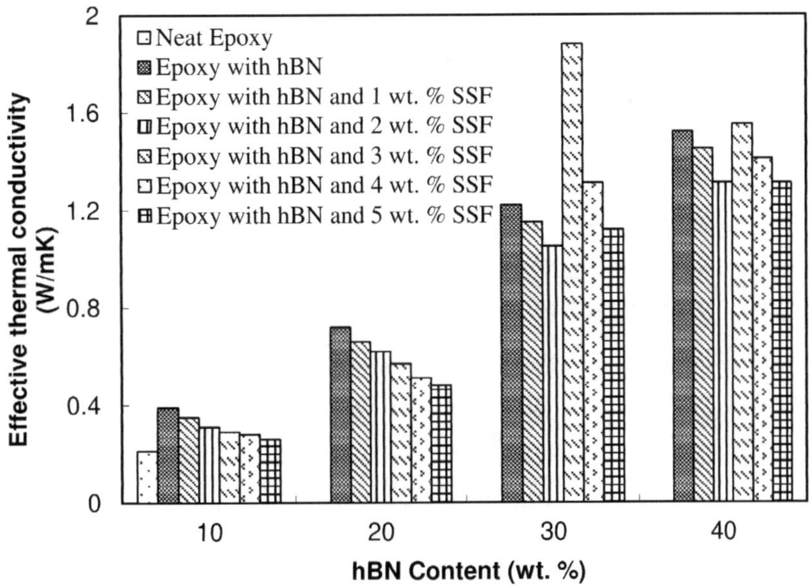

Figure 3. Thermal conductivity of fabricated samples.

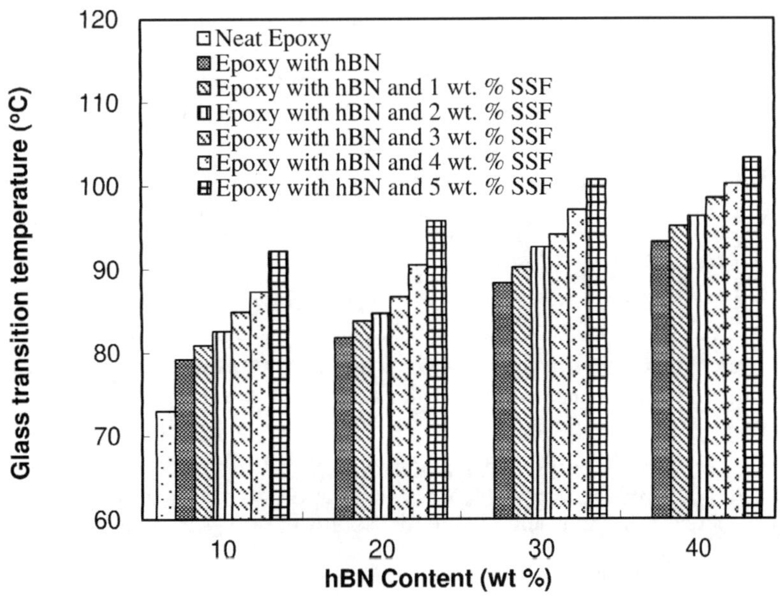

Figure 4. Glass transition temperature of fabricated samples.

IOP Conf. Series: Materials Science and Engineering **840** (2020) 012011 doi:10.1088/1757-899X/840/1/012011

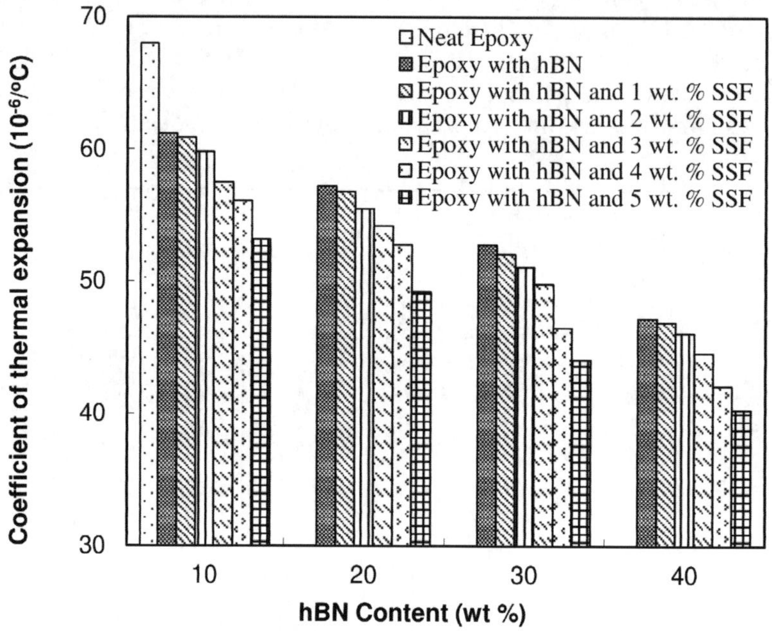

Figure 5. Coefficient of thermal expansion of fabricated samples.

3.4. Glass transition temperature

The variation in glass transition temperature of epoxy composites filled with different content of hexagonal boron nitride and sisal fiber is shown in figure 4. It is clear from the figure that compared to neat epoxy, glass transition temperature of its composites were higher. It can also be seen that the glass transition temperature depends upon the content of the filler. With single filler hBN, glass transition temperature increases from 73 °C for unfilled epoxy to 93.2 °C for 40 wt. % filled epoxy. Further, with incorporation of sisal fiber in the combination with hBN, glass transition temperature further increases and reaches maximum to 103.3 °C for a reinforcement of 40 wt. % hBN and 5 wt. % sisal fiber. The increment in glass transition temperature with the addition of the filler is because of the reduction in free volume of the polymer. With reduces availability of free volume, mobility of polymer chain is restricted and because of this restriction glass transition temperature of polymer composite increases [16].

3.5. Coefficient of thermal expansion

The variation in coefficient of thermal expansion of epoxy composites filled with different content of hexagonal boron nitride and sisal fiber is shown in figure 5. CTE value decreases with addition of either of the fillers into epoxy matrix. The decrement is attributed to constrained mobility which arises in the molecules of epoxy due to adsorption of filler surfaces. The CTE of the composite reduces from 68 x 10-6/°C to 40.3 x 10-6/°C with 40 wt. % of hBN and 5 wt. % of sisal fiber. The decreasing trend is gradual with filler content. With increase in content of either of the filler, there is a transition from loosely bound polymer to tightly bound polymer because of the reduction in the distance between filler content. This increase the hindrance inside the multi-phase system and supresses the expansion of polymer-filler combination.

4. Conclusions

The effect of micro-size hBN and short sisal fiber content on mechanical and thermal behaviour of epoxy based hybrid composites is investigated. Mechanical property analysis of epoxy/hBN/sisal fiber composites revealed that tensile strength increased with hBN content when 10 wt. % of hBN is added.

However further increase in hBN content decreases the tensile strength of the composite. Again, when combination of sisal fiber and hBN is incorporated, it successfully enhances the tensile strength of epoxy matrix but only when the content of sisal fiber is limited. With increase in either of filler content, compressive strength of the composites increases. Compressive strength increases from 85 MPa for neat epoxy to 108.7 MPa for 40 wt. % hBN in epoxy. It further increases to 118.1 MPa when 5 wt. % sisal fiber were added in combination with hBN in epoxy matrix. Thermal property analysis reflects that with increase in content of either hBN or sisal fiber, coefficient of thermal expansion and glass transition temperature increases and the increment is a function of filler content. Also, for a particular combination of filler, even insulative sisal fiber in combination with hBN help to increase the thermal conduction behavior of the epoxy matrix.

5. References

[1] Kim YK, Chung JY, Lee JG, Baek YK and Shin PW 2017 *Composites Part A: Applied Science and Manufacturing* **98** 184-91.
[2] Hou J, Li G, Yang N, Qin L, Grami ME, Zhang Q and Qu X 2014 *Rsc Advances* **4** 44282-90.
[3] Pan C, Kou K, Jia, Q, Zhang Y, Wu G and Ji T 2017 *Composites Part B: Engineering* **111** 83-90.
[4] Shen D. Zhan Z, Liu Z, Cao Y, Zhou L, Liu Y and Jiang N 2017 *Scientific reports* **7** 2606-17.
[5] Wang Z, Cheng Y, Wang H, Yang M, Shao Y, Chen X and Tanaka T 2017 *Journal of materials science* **52** 4299-308.
[6] Lule Z and Kim J 2019 *Polymers **11*** 148-63.
[7] Jiang Y, Shi X, Feng Y, Li S, Zhou X and Xie X 2018 *Composites Part A: Applied Science and Manufacturing* **107** 657-64.
[8] Jia ZR, Gao ZG, Cheng YH, Wu GL and Wu HJ 2018 *Chinese Physics B* **27** 117806-14.
[9] Agrawal A and Chandrakar S 2019 *Polymer Composites.*
[10] Kumar S, Joshi A and Gangil B 2013 *Proceedings of ICETED, AETS*, 200.
[11] Jagannatha TD and Harish G 2015 *International Journal of Mechanical Engineering and Robotics Research* **4** 131-137.
[12] Prasath KA, Krishnan BR and Arun CK 2013 *International. Journal of Mechanical Engineering* **2** 279-290.
[13] Ramesh M, Palanikumar K and Reddy K H 2013 *Composites Part B: Engineering* **48** 1-9.
[14] Kim JT and Netravali AN 2010 *Composites Part A: Applied science and manufacturing* **41** 1245-52.
[15] Xia C, Garcia AC, Shi SQ, Qiu Y, Warner N, Wu Y and Nie X 2016 *Scientific reports, 6,* 34726-34.
[16] Yung KC, Zhu BL, Yue TM and Xie CS 2009 *Composites Science and Technology* **69** 260-64.

Acknowledgments
This work was supported by the grant from the Science and Engineering Research Board, India (File No.: ECR/2017/000540).

Mechanical Behavior Analysis of Stainless Steels Subjected to Uniaxial Stress Tests

J Brnic[1,*], S Krscanski[1] and M Brcic[1]

[1] University of Rijeka, Faculty of Engineering, Department of Engineering Mechanics, Vukovarska 58, 51000 Rijeka, Croatia

[*]Email: brnic@riteh.hr

Abstract. The fundamental problem considered in this paper is the choice of materials, in this case stainless steels, for a particular industrial purpose. The choice of the appropriate material should be made based on the material behavior database established by experimental investigations. The mechanical behavior and mechanical properties of several stainless steels, such as ferritic, martensitic and austenitic stainless steels are investigated and analyzed in this research. Mechanical properties at room and high temperatures, creep behavior at different high temperatures and different stress levels as well as Charpy fracture impact energy were considered for several of the steels mentioned. Fracture toughness, based on the Charpy impact energy is calculated using known analytical procedure. In this sense, material properties are determined based on displayed engineering stress-strain diagrams while material creep behavior is displayed in the form of creep curves. Some of numerical values regarding the ultimate tensile strength, measured at room temperature, are given using the following order and in the following form: $[\sigma_{m,Mat.Nr}]_{20°C}$. In this sense, the stress levels of the u are: $[925_{1.4122}; 782_{1.4034}; 726_{1.4305}; 686_{1.4841}; 585_{1.4762}; 607_{1.4541}]_{20°C}$.

1. Introduction

Modern design of engineering structures or their elements is based on proper selection of materials and numerical control of mechanical behavior. In any case, designing of structures that must be safe, reliable and economical, requires efficient selection and use of material but also assurance that failure during structure life will not occur [1]. Over the long term, the finite element method has become a very attractive technique in the computer processing of complex engineering solutions [2]. In accordance with the purpose of the structure, some structure is subjected to high stress, another one to high temperature or to repeated load, etc. However, design of the considered structure need to be made from the best material that can meet all of requested requirements, and usually, under assumption that material is free of any failure and that no failure during structure life will occur. In spite of this desire and all the best design pleasures, various failures are possible during the operation of the structure under the prescribed environmental conditions and loads. Usually, in engineering practice, two main groups of failures can be spotted and those are preexisting failures in material and failures that can occur during structure lifetime (from design through maintenance and inspection to the regular workload). Commonly observed mechanical failures can be listed as excess deformation, corrosion, impact, creep, fracture, fatigue, wear, etc. [3]. In this investigation, apart the mechanical properties as the most important task at material selection, some of mentioned failures, such as yielding, creep or, for example, fracture impact energy, were explored. It is known that, in general, creep can occur in metals at temperature of 370°C, or it can be said that in engineering practice creep is appreciable above $\mathbf{0.4}\boldsymbol{T_m}$, where $\boldsymbol{T_m}$ is melting point measured in Kelvin [4]. In this sense the investigation of creep behavior of considered materials is very important. From the other hand, investigation regarding the energy of impact fracture is also of great importance, since based on it, someone can assess the fracture toughness using, for example known analytical formula. The main intention in these investigations was to provide information in one place of utmost importance for the selection of materials and the possibility of their comparison. With regard to the materials being researched, some of the results by individual research items can be found in the available literature, such as the following papers. In Refs. [5-6], some investigation results regarding the material properties and mechanical behavior of material 1.4762 are presented, in Refs. [7-8] also some results regarding mechanical properties and corrosion behavior of 1.4122 material are presented. Further, in Refs. [9-10], experimental results regarding material properties of material 1.4841 are given. Similar, in Refs. [11-12] experimental results regarding the mechanical behavior of material 1.4541 are presented. In addition, for other here investigated materials, also some data regarding their mechanical behavior can be found in the published papers. Experimental tests were carried out according to the standards given in Ref. [13].

1.1 *The significance of this research*
This research has a wide range of meanings. The study of mechanical properties indicates the level of stress and deformation that the studied materials can withstand at different temperatures. Furthermore, the creep resistance indicates at which temperatures and load levels the materials can be used for a certain period of time, and finally what the impact energy of the fracture of individual materials is. Taken together, the result indicates how well these materials are suitable for particular applications.

2. Materials, equipment and standards
Materials whose mechanical behaviors are being investigated are (mat. numbers): 1.4122, 1.4034, 1.4305, 1.4841, 1.4762 and 1.4541. Their chemical compositions regarding to only some of constituents are presented in table 1, while in table 2 are given data regarding the type of the material and application possibilities.

Table 1. Tested materials: Chemical compositions (some elements only), mass %

Material / round bar (mm)	Chemical composition (some constituents only), mass %					
	C	Cr	Si	Ni	Mn	Mo
1.4122 (X39CrMo17-1) /18	0.39	16.4	0.45	0.35	0.56	0.81
1.4034 (X46Cr13) /16	0.44	13	0.37	-	0.38	-
1.4305 (X10CrNiSi18-9) /18	0.047	17.3	0.58	8	2	-
1.4841 (X15CrNiSi25-20) /16	0.027	24	2.7	20.2	1.6	0.2
1.4762 (X10CrAlSi25) /18	0.102	23	1.2	0.68	0.52	0.12
1.4541 (X6CrNiTi18-10 /18	0.017	17	0.44	9.3	1.44	0.24

Table 2. Tested materials: Type and application possibility (A-austenitic; F-ferritic; M-martensitic)

Material	Type	As-received	Application
1.4122 (X39CrMo17-1	M	-	Mechanical engineering (compressor and pumping parts, shafts, valves, polymer industry)
1.4034 (X46Cr13)	M	annealed+cold drawn	Mechanical-civil-industrial engineering (shafts, pumps, gears, etc)
1.4305 (X10CrNiSi18-9)	A	cold drawn	Mechanical engineering (heavily machined parts, shafts, gears valve parts,screws, bolts, food ind.)
1.4841 (X15CrNiSi25-20)	M	annealed+cold drawn	Aerospace – shipbuilding - military industry, nuclear power industry. Thermal treatment plants.
1.4762 (X10CrAlSi25)	F	annealed	Oil-chemical-petroleum-ceramics-cement industry, furnace, steam boilers technology.
1.4541 (X6CrNiTi18-10	A	annealed	Automotive-chemical-building-construction-aerosoace industry.

3. Test results and analysis

3.1 Mechanical behavior of materials at different temperatures

Engineering stress-strain diagrams for tested materials are shown in figure 1. Based on these figures it is possible to make some comparison regarding the ultimate tensile strength. See also data in table 3.

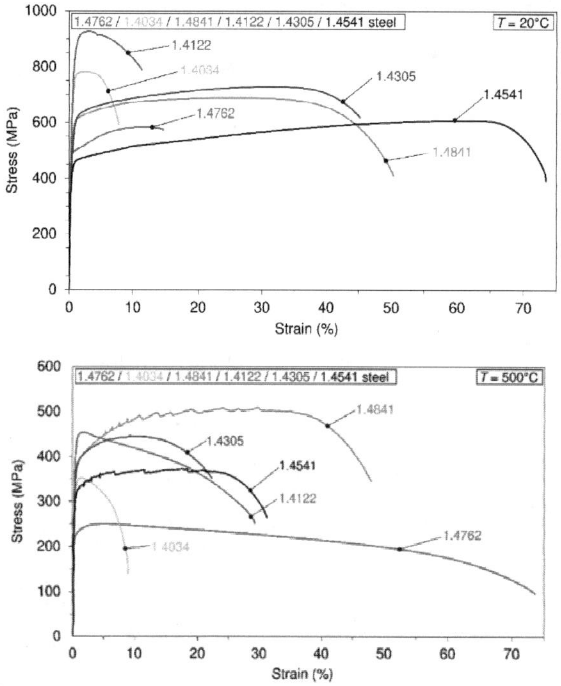

Figure 1. Engineering stress-strain diagrams of tested material at room temperature and that of 500°C

3.2. Short-time creep tests

Short-time creep tests for considered materials are shown in figure 2 regarding the temperatures of 500°C and 600°C. At temperature of 500°C all of tested materials were subjected to stress level of 0.5 yield strength at considered temperature. Although stress level is not the same for all of materials, it is visible that materials 1.4841, 1.4305, 1.4541 may be treated as creep resistant at these conditions, while other considered materials may not be. For the same materials tested at temperature of 600°C and stress levels shown in figure 2, it can also be said that they are creep resistant.

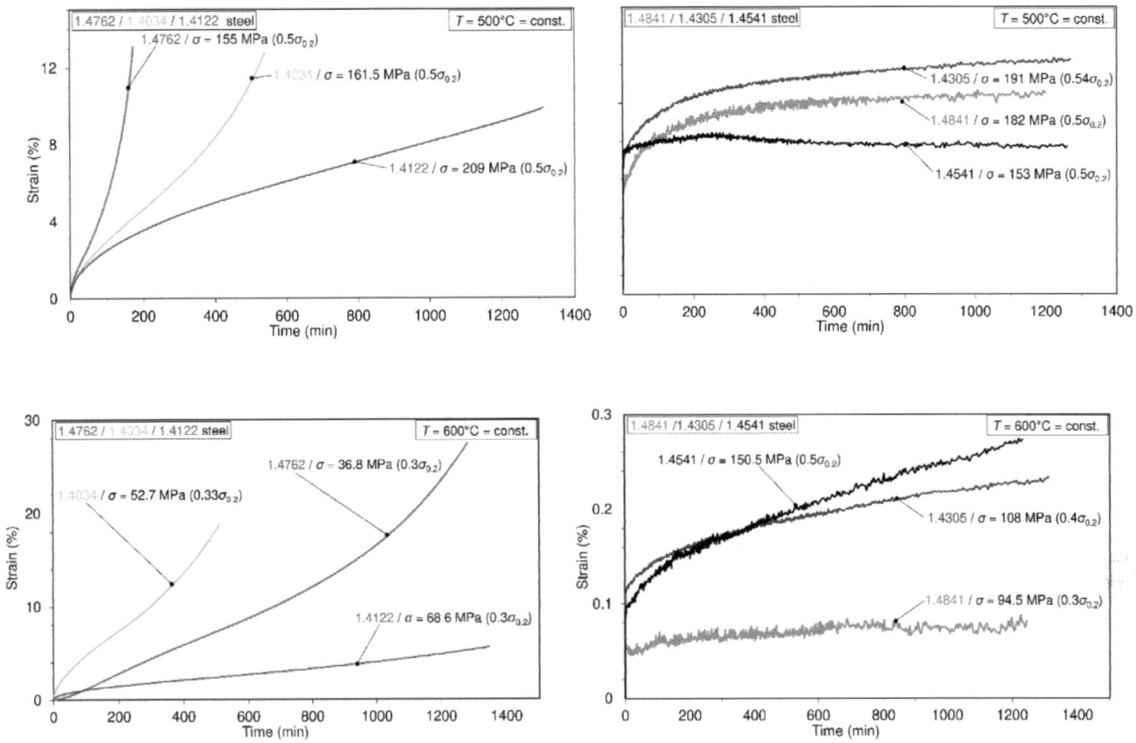

Figure 2. Creep curves of tested materials at temperatures of 500°C and 600°C

The effect of temperature on considered materials may be summarized as it is shown in table 3.

Table 3. Effect of temperature on mechanical behavior of considered materials

Material	Type	Ultim. tensile strength, σ_m MPa		$\sigma_{m,500}/\sigma_{m,20}$	Creep resistant 500°C, 600°C
		20°C	500°C	ξ	+/−
1.4122	M	925	453	0.49	−
1.4034	M	782	352	0.48	−
1.4305	A	726	445	**0.61**	+
1.4841	A	686	508	**0.74**	+
1.4762	F	585	270	0.46	−
1.4541	A	607	372	**0.61**	+

3.3 Charpy V-notch impact fracture energy

In figure 3 curves representing Charpy impact energy are shown. Using well known Roberts-Newton formula, fracture toughness can be calculated. The result obtained is independent of temperature.

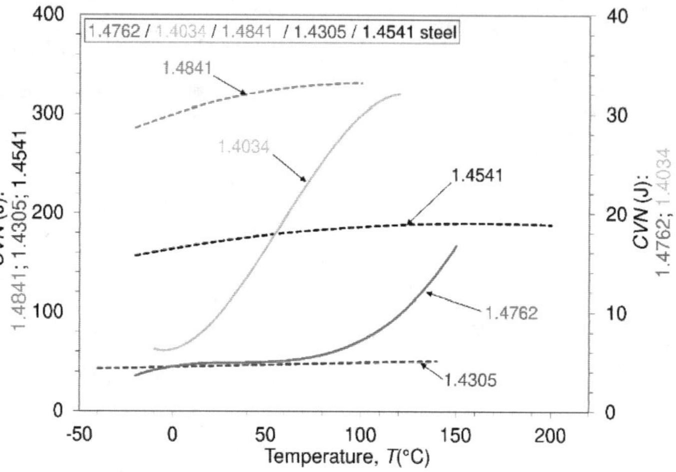

Figure 3. Charpy V-notch impact fracture energy of tested materials.

In accordance with the results of impact fracture energy shown in figure 3, it is visible that the highest value of it occurs at the steels 1.4841 and 1.4541.

4. Conclusion

In this investigation very important and useful experimental data regarding mechanical properties, creep behavior as well as Charpy impact fracture energy were obtained. Based on these results it is visible that the highest ultimate tensile strength at room temperature possesses steel 1.4122, while at temperature of 500°C, the steel 1.4841. Furthermore, the considered steels 1.4305, 1.4841 and 1.4541 can be treated as creep resistant at the temperatures of 500°C and 600°C and for the applied stress levels, while this does not apply to other materials tested. The relatively smallest influence of temperature is evident in the mechanical behavior of the materials 1.4305, 1.4841 and 1.4541.

5. References

[1] Dowling N E 2013 *Mechanical Behavior of Materials* (New York: Pearson)

[2] Bathe K J 1982 Finite Element Procedures in Engineering Analysis (New Jersey: Prentice- Hall, Inc)

[3] Stephens R I, Fatemi A, Stephens R R and Fuchs H O 2001 *Metal Fatigue in Engineering* (New York: John Wiley & Sons)

[4] Raghavan V 2004 *Materials Science and Engineering* (New Delhi: Prentice- Hall of India).

[5] Brnic J, Turkalj G, Krscanski S, Lanc D, Canadija M and Brcic M 2014 Information relevant for the design of structure: ferritic – heat resistant high chromium steel X10CrAlSi25 *Materials and Design* **63** 508-518.

[6] Krumla T and Polákab J 2001 Fatigue softening of X10CrAl24 ferritic steel, *Materials Science and Engineering: A* **319–321** 564-568.

[7] Brnic J, Turkalj G, Vukelic G and Brcic M 2012 Analysis of the dependence of material properties on temperature – steel 1.4122 *High Temperature Materials and Processes* **31** 259–266.

[8] Ensling J, Fleisch J, Grimm R, Grüber J and Gütlich P 1978 *Corrosion Science* **18** 9 797-804.

[9] Brnic J, Turkalj G, Canadija M, Krscanski S, Brcic M and Lanc D 2015 Deformation behaviour and material properties of austenitic heat-resistant steel X15CrNiSi25-20 subjected to high temperatures and creep *Materials & Design* **69** 219-229.

[10] Caty O, .Maire E, .Douillard T, .Bertino P, .Dejaeger R and .Bouchet R 2009 Experimental determination of the macroscopic fatigue properties of metal hollow sphere structures, *Materials Letters* **63** 13–14, 1131-1134.

[11] Brnic J, Turkalj G, Canadija M, Lanc D, Krscanski K, Brcic M, Li Q and Jitai Niu 2016 Mechanical properties, short time creep, and fatigue of an austenitic steel, *Materials* **9** 298 doi:10.3390/ma9040298.

[12] Celik O, Baydogan M, Atar E, Kayali S and Cimenoglu H 2013 Fatigue performance of low temperature nitride AISI 321 grade austenitic stainless steel *Materials Science and Engineering A* **565** 38–43.

[13] *Metallography: Metals–Mechanical Testing: Elevated and Low–Temperature Tests (Annual Book of ASTM Standards vol. 03.01)* 2015 (ASTM International: Baltimore).

Acknowledgments

This work has been fully supported-supported in part by Croatian Science Foundation under the project **IP-2019-04-8615,** as well as by the University of Rijeka within the project: **uniri-technic-18 - 42.** The authors (Josip Brnic, Sanjin Krscanski and Marino Brcic) are grateful for the financial support.

IOP Conf. Series: Materials Science and Engineering **840** (2020) 012013 doi:10.1088/1757-899X/840/1/012013

Morphological and Mechanical Properties of Natural Rubber Compound/Poly(butylene succinate) Blend

N Prasoetsopha[1*], P Thainoi[1], R Jinnavat[1], W Charerntanom[1], A Hasook[1] and W Singsang[2]

[1] Department of Materials Engineering, Faculty of Engineering and Architecture, Rajamangala University of Technology Isan, Sub-district Nai Muang, District Muang, Province Nakhon Ratchasima 30000, Thailand
[2] Department of Industrial Technology, Faculty of Industrial Technology, Rambhai Barni Rajabhat University, Sub-district Ta-Chang, District Muang, Province Chanthaburi 22000, Thailand

*Email : natkrita.pr@rmuti.ac.th

Abstract. Biodegradable poly(butylene succinate) (PBS)/natural rubber compound (NRC) blend was melt mixed using a two-roll mill. A weight ratio of PBS/NRC was 100/0, 90/10, 80/20, 70/30, 60/40 and 50/50. Morphological and mechanical properties were investigated. The results showed that the elongation at break increased with an increase of NRC content until the ratio of 80/20. While the tensile strength continuously decreased with an increasing amount of NRC. Moreover, the impact strength of the PBS/NRC blend increased until the ratio of 80/20 and tended to decrease in the further ratios. Morphology of PBS/NRC blend showed that NRC had good dispersion in PBS matrix up to 80/20. When the amount of NRC was increased, it became a larger particle due to an agglomeration, resulting in the mechanical properties.

1. Introduction

Thermoplastic elastomer (TPE) is a one of the polymer blends with a combination of rubber and thermoplastic. The category of TPE can be divided into two groups namely simple blends (SBs) and thermoplastic vulcanizes (TPVs). The advantages of blending elastomer and thermoplastic are that they are easy to process in melt mixing and creating a new type of materials with more anticipated properties. Hence, there are many attractive properties of the thermoplastic elastomer [1-6]. Typically, the final morphology of thermoplastic elastomer is the dispersion of rubber particles in thermoplastics matrix.

Natural rubber (NR) is a bio-based elastomer. It is has a high elasticity and good tear strength. There are many attempts to improve the toughness of polymeric matrix by natural rubber [2-5]. The example of some studies were reported in polystyrene [2], poly(lactic acid) [3-4], poly(butylene succinate) [5-6].

Nowadays, bioplastics have attracted a considerable interest to study and research on the previous report [1, 3-6]. The raw materials which make bioplastics are renewable resources such as cellulose, casein, collagen, soy protein, corn. This is the best way to decrease the plastic pollution because bioplastics can degrade and decompose in the natural environment. The examples of bioplastics are polycaprolactone (PCL), poly(lactic acid) (PLA), and poly(butylene succinate) (PBS). The most popular material in the bioplastic industry is PLA. However, it is brittle material. On the other hand,

IOP Conf. Series: Materials Science and Engineering **840** (2020) 012013 doi:10.1088/1757-899X/840/1/012013

PBS has good mechanical properties and is easy to process. PBS is synthesized from the condensation reaction between 1,4-butandiol and succinic acid. It has a polyester and linear structure. It has thermal stability up to 200°C. After use, PBS degrades into carbon dioxide and water. However, it is still an expensive cost. Hence, there are many attempts to increase the ability application of PBS i.e. polymer blends [5-7] and/or composites [8-9].

In this research, the biodegradable PBS and natural rubber compound (NRC) in an effective vulcanization system was melt blended in a two-roll mill. The effect of NRC content on the morphological and mechanical properties was investigated.

2. Materials and method

2.1. Materials

Natural rubber (NR, STR 5L), steric acid, zinc oxide (ZnO), tetramethyl thiuram disulphide (TMTD), N-cyclohexyl-2-benzothiazole sulfenamide (CBS), and sulphur (S) were used to prepare natural rubber compound (NRC) in an effective vulcanization system. All chemicals were purchased form Chareon Tut Co., Ltd., Thailand. Biodegradable PBS, BioPBSTM FZ71PM grade, was supplied by PTT MCC Biochem Co., Ltd.

2.2. Natural rubber compound preparation

Natural rubber (NR) and chemicals were mixed via a two-roll mill (Chareon Tut co., ltd., ML-D6L12-INV) with a roller speed of 10 rpm for 30 min in order to make NRC. The order of reagents is shown in Table 1. After being mixed, the NRC was kept in room temperature for a night before being blended.

Table 1. Natural rubber compound formula.

No.	Raw materials	Composition, phr	Mixing time, min
1.	Natural rubber	100	5
2.	Steric acid	2	5
3.	ZnO	6	5
4.	TMTD	3.5	5
5.	Sulphur	0.75	5
6.	CBS	1	5

2.3. Poly (butylene succinate)/natural rubber compound blend

PBS was dried in an oven at 50°C for 24 h before use. NRC and dried PBS were mixed with a two-roll mill plastic (Chareon Tut co., ltd., ML-D6L12HT-INV) at 120°C for 20 min. The ratio of PBS/NRC was 100/0, 90/10, 80/20, 70/30, 60/40, and 50/50. The total amount of each batch was 500 g. After completely being mixed, the PBS/NRC blend was cooled to room temperature and ground with a grinder.

2.4. Hot-compression process

The ground PBS/NRC blends were dried in an oven at 50°C for 24 h before hot-compression process. Hot-compression machine (Chareon Tut co., ltd., PR2D-W300L350 PM-WCL-HMI) was used to prepare samples at a temperature of 145°C and pressure of 1500 psi in order to investigate the mechanical properties. The pre-heat time, pressing time and cooling time were 5, 5, and 3 min, respectively. The sample for mechanical testing was prepared according to ASTM D638 for tensile test and ASTM D256 for impact test.

IOP Conf. Series: Materials Science and Engineering **840** (2020) 012013 doi:10.1088/1757-899X/840/1/012013

2.5. Testing and characterization

2.5.1. Mechanical testing. For impact test, the notched Izod impact type was conducted according to ASTM D256 at room temperature by impact tester machine (INSTRON, Ceast 9050). At least ten samples were tested and evaluated. For tensile test, the universal testing machine (UTM, LLOYD INSTRUMENT, LR10K plus) was used with a crosshead speed of 50 mm/min at room temperature.

2.5.2. Morphological study. The fracture surface from impact sample was investigated by using scanning electron microscope (JEOL, JSM-6510).

3. Result and discussions

3.1. Mechanical testing.
In this research, the mechanical properties of poly(butylene succinate)/natural rubber compound (PBS/NRC) blends were investigated in terms of impact and tensile tests.

For impact test, the impact strength of PBS/NRC blends is illustrated in Figure 1. The results showed that the impact strength of neat PBS was 6.2 ± 0.6 kJ/m^2. After NRC blending, the PBS/NRC blends in the ratio of 90/10, 80/20, 70/30, 60/40, and 50/50 showed the impact strength of 8.61 ± 1.0, 12.09 ± 1.4, 11.3 ± 1.2, 9.2 ± 1.2, and 7.8 ± 0.8 kJ/m^2, respectively. The highest impact strength was the PBS/NRC blends of 80/20. This was a 2-fold enhancement compared with the neat PBS. The reason of this improvement was that the NRC acted as stress concentration in polymeric matrix. It is a toughened polymer matrix similar with a previous report [6].

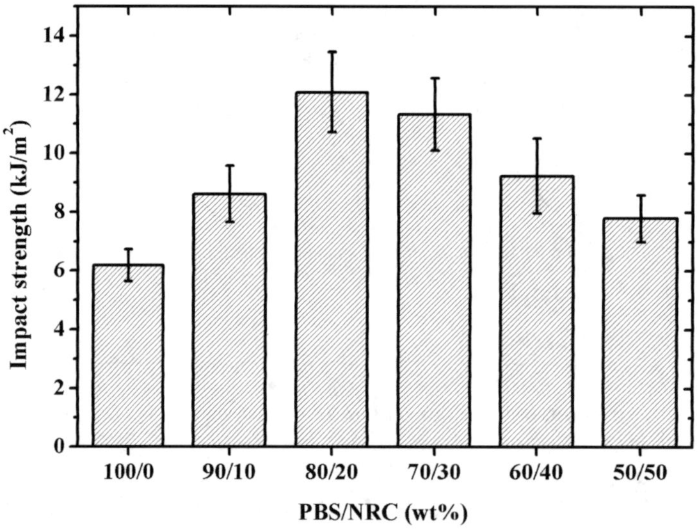

Figure 1. Impact strength of poly(butylene succinate)/natural rubber compound (PBS/NRC) blends

For tensile test, the elongation at break and tensile strength are illustrated in Figure 2. The elongation at break of PBS/NRC blends showed that the NRC contributed to improve this value (Figure 2a). The highest elongation at break of the blends was 29.1 ± 4.6 percentage in the ratio of 80/20. It was enhanced to 1.7 times compared with the neat PBS. This is due to the elastic property of NRC similar to the polymer-rubber system [6]. On the other hand, the incorporation of NRC significantly decreased the tensile strength of PBS (Figure 2b). As expected, tensile strength decreased with an increasing amount of NRC.

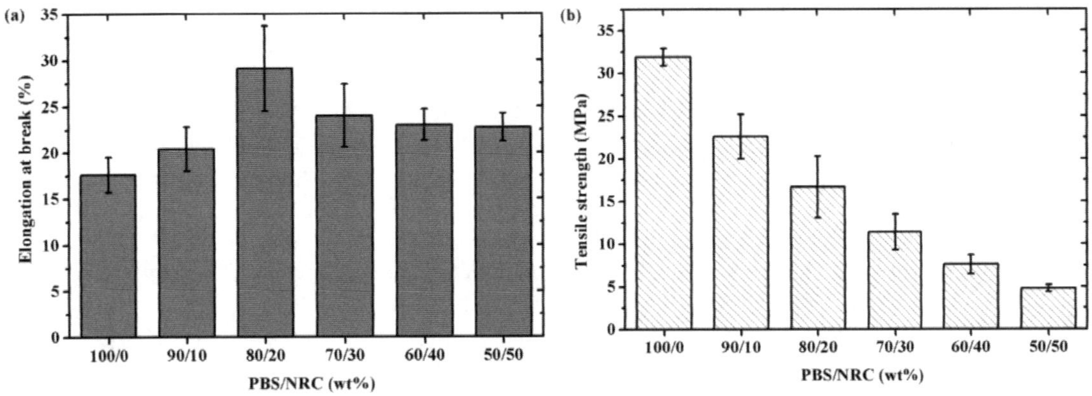

Figure 2. Tensile properties of poly(butylene succinate)/natural rubber compound (PBS/NRC) blends; (a) elongation at break and (b) tensile strength

3.2. Morphological study.

Figure 3 illustrates the SEM micrographs of the fractured surface of PBS/NRC blends with various blend ratios. The result showed that the coarse surface of pure PBS could be observed (Figure 3a). All blends illustrated the dispersed NRC particles in the PBS matrix (Figure 3b-f). Especially, the blends with 10 and 20 wt% of NRC exhibited the dispersed particle size of 2-5 μm as shown in Figure 4b-c. When the blends had a higher NRC content, the large dispersed particle size occurred due to the agglomeration of rubber particles as seen in Figure 3d-f. These morphologies were involved with the mechanical properties as previous discussion. From the mechanical and morphological properties, the PBS/NRC of 80/20 was the best blend ratio.

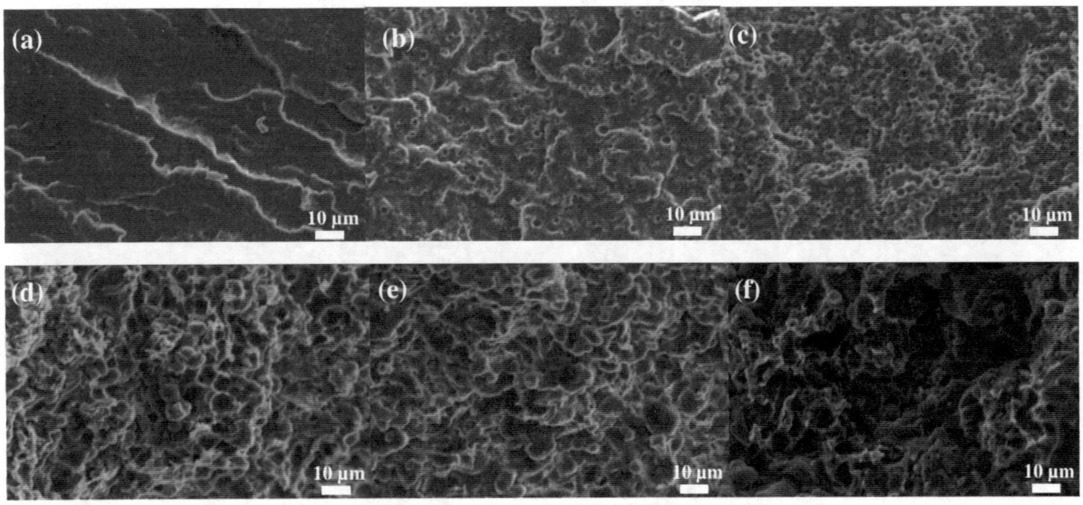

Figure 3. Morphology of PBS/NRC blends with a weight ratio of; (a) 100/0, (b) 90/10, (c) 80/20, (d) 70/30, (e) 60/40 and (f) 50/50

4. Conclusions

PBS/NRC blends with various weight ratios of 100/0, 90/10, 80/20, 70/30, 60/40 and 50/50 were prepared via a two-roll mill. The amount of NRC had an effect on the mechanical and morphological properties of the blends. The highest impact strength and percentage elongation at break were achieved at 20 wt% of NRC. The tensile strength decreased monotonically with increasing NRC content. Finally, NRC might be used to enhance toughness of other brittle polymer matrix.

IOP Conf. Series: Materials Science and Engineering **840** (2020) 012013 doi:10.1088/1757-899X/840/1/012013

5. References

[1] Maccaferri E, Mazzocchetti L, Benelli T, Brugo T M, Zucchelli A and Giorgini L 2020 *Mater. Design.* **186** 108210
[2] Asletha R, Kumaran M G and Thomas S 1999 *Eur. Poly. J.* **35** 253-271
[3] Pongtanayut K, Thongpin C and Santawitee O 2013 *Energy Procedia* **34** 888-897
[4] Dechatiwon Na-Ayutthaya W and Poompradub S 2014 *Macromol. Res.* **22** 686-692
[5] Faibunchan P, Nakaramontri Y, Chueangchayaphan W, Pichaiyut S, Kummerlöwe C, Vennemann N and Nakason C 2018 *J. Poly. Environ.* **26** 2867-2880
[6] Faibunchan P, Pichaiyut S, Chueangchayaphan W, Kummerlöwe C, Vennemann N and Nakason C 2019 *Poly. Adv. Technol.* **30** 1010-1026
[7] Hemsri S, Thongpin C, Moradokpermpoon N, Niramon P and Suppaso M 2015 *Macromol. Symp.* **345** 145-154
[8] Huang Z, Qian L, Yin Q, Yu N, Liu T and Tian D 2018 *Polym.Test.* **66** 319-326
[9] Calabia B P, Ninoomiya F, Yagi H, Oishi A, Taguchi K, Kuioka M and Funabashi M 2013 *Polymers* **5** 128-141

Acknowledgments

This research project was supported by Rajamangala University of Technology Isan. Contract No. NKR2561REV033. The author would like to thank the staffs in Faculty of Engineering and Architecture and Institute of Research and Development for all kindly supports.

IOP Conf. Series: Materials Science and Engineering **840** (2020) 012014 doi:10.1088/1757-899X/840/1/012014

Study on the Influence of Blast Furnace Slag on Chloride Ion Penetration Property

P Subpa-asa[1], K Ken[2], F Satoshi[3], D Shigeyuki[4]

[1] Graduate student, Department of Civil Engineering, Graduate School of Science and Technology, Tokai University, Hiratsuka-pref., Japan, 259-1292.
[2] Chief manager, DC CO., LTD., Technical Center, 1-17, Asano-cho, Kawasaki-ku, Kawasaki-shi, Kanagawa, 210-0854, Japan
[3] Assistant manager, DC CO., LTD., Technical Center, 1-17, Asano-cho, Kawasaki-ku, Kawasaki-shi, Kanagawa, 210-0854, Japan
[4] Professor, Civil Engineering Department, Faculty of Engineering, Tokai University, Kanagawa-pref., Japan, 259-1292.

Email: prangsub@gmail.com

Abstract. Sustainability is becoming more and more important, which is why the concrete and cement industry are attempting to reduce negative impacts to the environment. One of the methods regarding this is to use industry by-products from the manufacturing of steel products like blast furnace slag (BFS) to replace the raw materials required in cement. The chloride ion penetration has to also be taken into consideration since it reinforces the concrete structure. Thus, the influence of BFS on blocking resistance and chloride ion penetration by using BFS with different Blaine value is studied by using "Standard on Test Methods for Chloride Ion Diffusion Coefficients in Concrete" by electrophoresis (Draft) (JSCE-G571-2003)" and the total chloride ion amount was measured in accordance to JIS R 5202. The result is that using BFS in cement has influences on the penetration properties.

1. Introduction

Reinforced concrete structures (RC structure) where steel is embedded and the two materials act together as resisting forces. Chloride ions are also infiltrated and corrosion of the reinforcing bars can occur, causing cracks. These cracks from corrosion expansion are the chloride damage which then affects the performance of the structures. Concrete which is a constituent material of the structure is required to have high durability to prolong its use.

Steel slag is the unavoidable by-product of the manufacturing of steel products which includes useful materials like blast furnace slag (BFS). Japan has accumulated approximately 40 million tons of iron and BFS products. The largest use of BFS is in cement, as raw materials. In addition, the total annual sales for domestic consumption and exports reach a combination of 17.8 million tons. BFS is also used as a fine aggregate for concrete, or as sand. Fine aggregate is the second-largest use of BFS, with sales of 1.8 million tons per year in Japan. [1] It has been found that concrete with BFS has excellent resistance to sulfuric acid, freezing and thawing resistance and resistance to damage by salt attacks. [2-4] Furthermore, BFS is more effective in preventing corrosion of the reinforcing bars in concrete due to its denser cured body, as compared with ordinary Portland cement, and its high ability to immobilize chloride ions. Although, the mechanism of immobility of chloride ions in concrete containing BFS has not been clarified.

IOP Conf. Series: Materials Science and Engineering **840** (2020) 012014 doi:10.1088/1757-899X/840/1/012014

Cement replacement levels were generally conserved at 30-70% by the fine BFS powder. The workability of cement was improved by the adjusting water-cement ratio. [5] To compare, when the fine BFS powder was added, the properties changed as air void, the long-term strength, the chemical resistance and density were improved. Therefore, this study is aimed to confirm the influence of BFS on blocking resistance and the performance of chloride ions by using three different Blaine values.

Table 1. Material properties.

Abbreviations	Properties	Density (g/cm^3)
N	Ordinary Portland cement	3.16
BFS	Ground granulated Blast Furnace slag (Blaine value of 3,000 4,000 6,000)	2.90
S	Sand	2.64
AG	Anhydrous Gypsum plaster	2.91
W	Water	1.00

2. Methodology

2.1 Materials and formulations used

Table 1 and Table 2 show materials properties and mixed proportions. The BFS and anhydrous gypsum plaster (AG), adjusted with the amount of SO3 were 2.41% (the amount of SO3 in level 1) the larger the amount SO3, the early strength increases and shrinkage decreases. The percentages of the BFS replacement used in this experiment were 45% and 22.5% and the latter will confirm the change in the performance of the chloride ions' immobility.

2.2 Mixing-curing method

The proportions of the mortar were mixed in accordance with JIS R5201," Physical testing methods of cement". [6] The curing condition was at two levels which are water curing and steam curing, the pre-time was 2 hours and the temperature was raised at 20℃/1hour in advance. Then, it was held at 60℃ for 24 hours. After that, it fell at 10℃/hour. Steam curing was carried out for 28 days.

2.3 Rapid chloride penetration test (RCPT)

The methodology of the chloride ion penetration test is shown in Fig. 1 and the specimen geometry has a diameter of 100 mm and a thickness of 50 mm in a cylinder shape. The influence of the BFS on blocking resistance and chloride ion penetration by using BFS with different Blaine values is studied by using the "Standard on Test Methods for Chloride Ion Diffusion Coefficients in Concrete" by electrophoresis (Draft) (JSCE-G571-2003) [7] and the total number of chloride ions was also measured.

Table 2. Mixed proportion.

Mix proportion	*W/B	Unit weight (kg/m³)						
		*W	N	BFS (Blaine value)			AG	S
				3,000	4,000	6,000		
Sample-1			636					
Sample-2			493	143			6.2	
Sample-3			350	268			12.4	
Sample-4	50%	318	493		143		6.2	1,272
Sample-5			350		268		12.4	
Sample-6			493			143	6.2	
Sample-7			350			268	12.4	

*W/B is the ratio of water to the binder

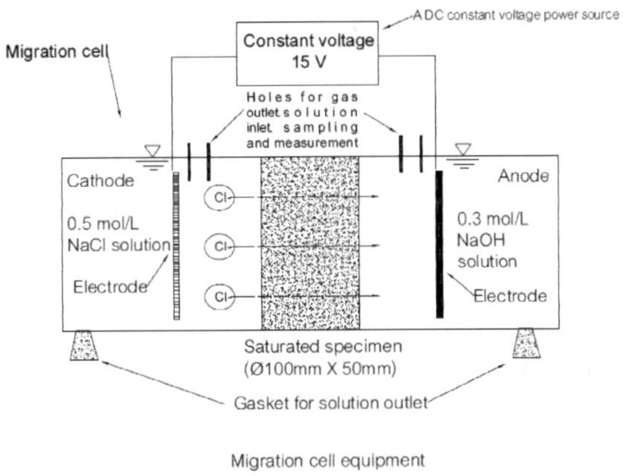

Figure 1. Chloride ion penetration test

3. Results and discussion

3.1 Chloride ions diffusion coefficient

Figures 2 and 3 show the results of the chloride ions diffusion coefficient of the specimen through steam curing and water curing. The chloride ions diffusion coefficient of concrete with no BFS has the same tendency with adding BFS by 22.5% of cement replacement, which is higher than adding BFS by 45% of cement replacement. The exception is that the penetration of chloride ions is suppressed even with the same amount of addition and there was a difference depending on the Blaine value. More resolution of the Blaine value affects the chloride ions diffusion coefficient in a way that it decreases. This could be because of the densification of the mortar as the degree of fineness increases. Products that went through steam curing had more texture than products that was cured by water, so the chloride ion property lowered.

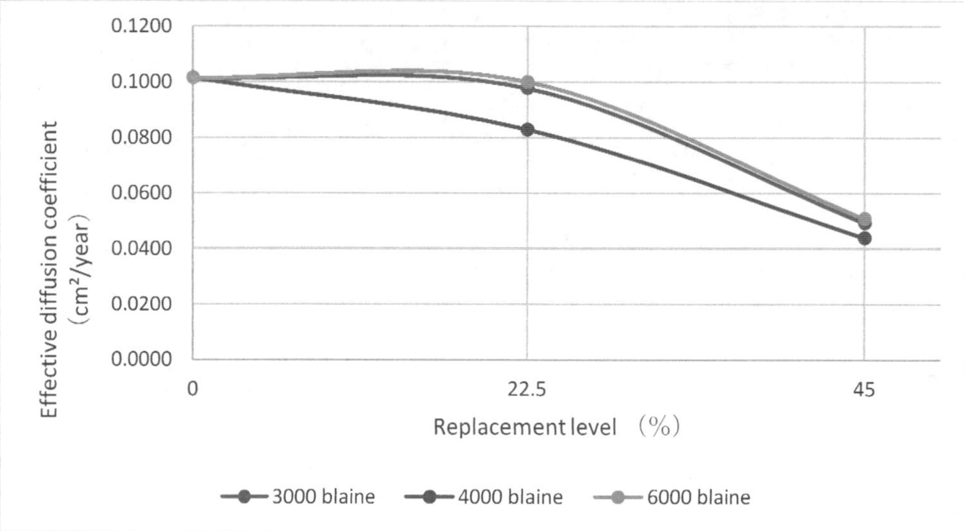

Figure 2. The result of the diffusion coefficient of the steam curing.

3.2 Chloride ion content

After confirming that the penetration rate of chloride ion is constant, the energization was stopped and the specimen was taken out. The amount of chloride ion penetration from the concrete was measured with a parameter analyzer. Figures 4 and 5 show the analysis results of water curing and steam curing, respectively. As a result, the addition rate and fineness of the ground granulated blast furnace slag increased, which causes the amount of chloride ions remaining in the specimen to increase.

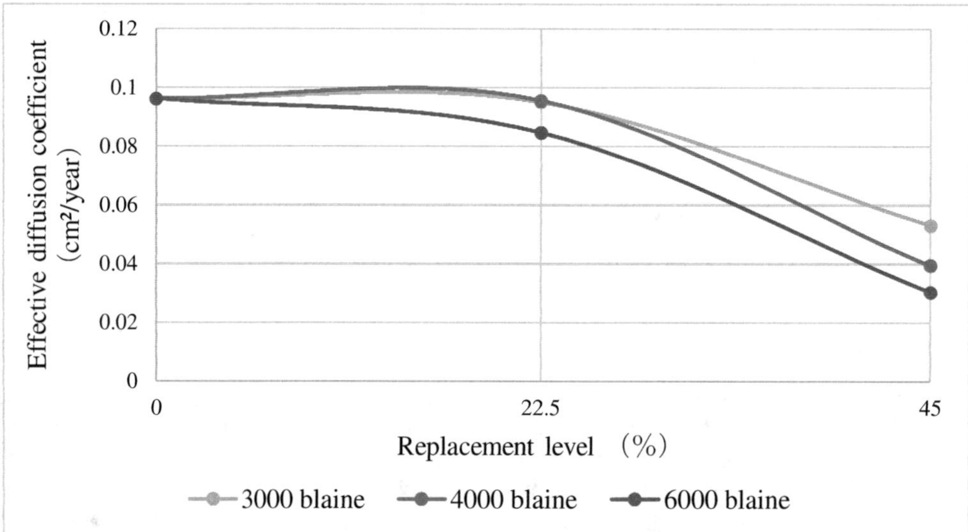

Figure 3. The result of the diffusion coefficient of the water curing.

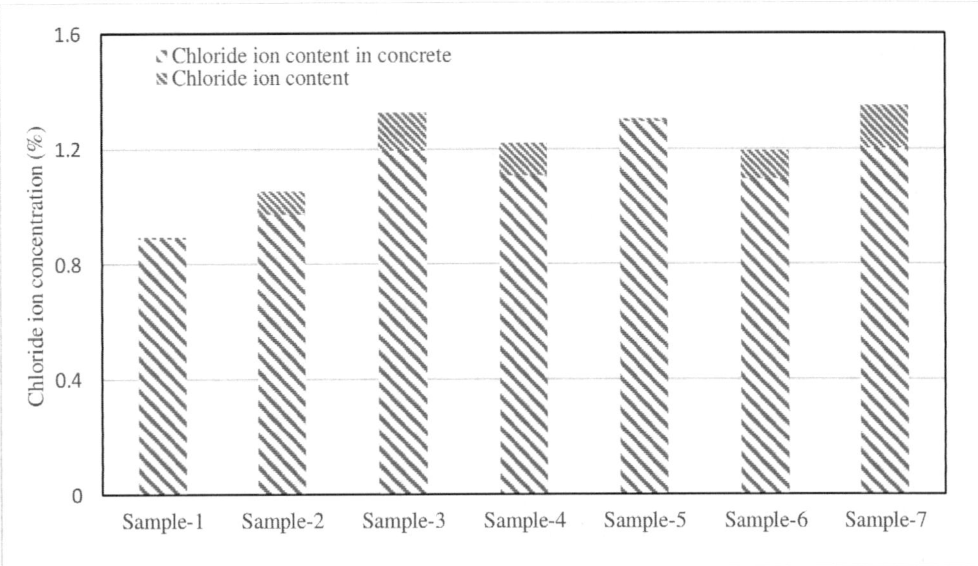

Figure 4. The analysis results of water curing.

The lead time of the BFS concrete was getting longer. The degradation prediction of the amount of chloride ions retained due to the chloride damage is currently based on the diffusion coefficient. On the other hand, the immobility of chloride ions leads to an increase in useful life from the viewpoint of corrosion of the reinforcement, so in the future, it would be beneficial to explore this effect further.

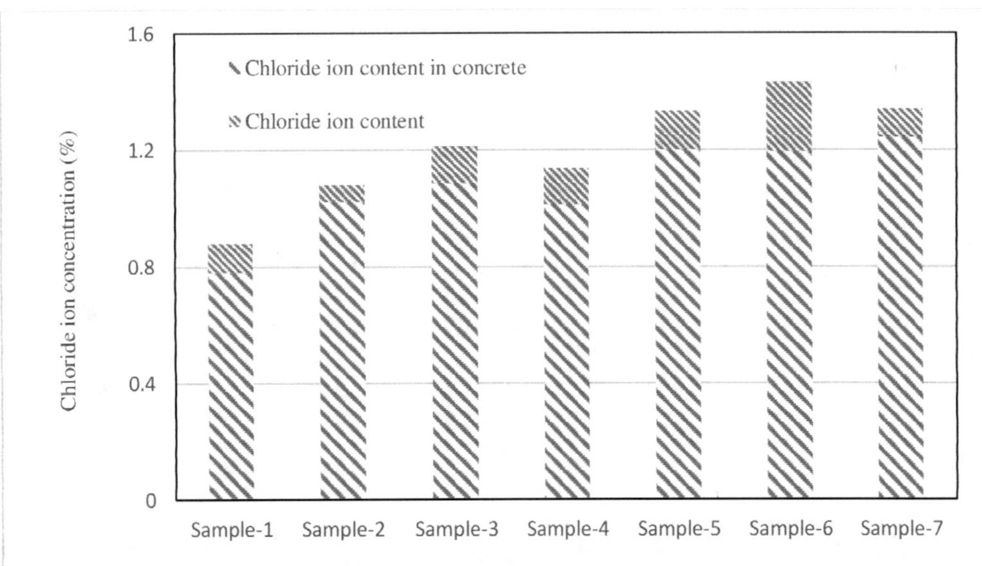

Figure 5. The analysis results of steam curing

IOP Conf. Series: Materials Science and Engineering **840** (2020) 012014 doi:10.1088/1757-899X/840/1/012014

4. Conclusion

The study investigated the possibility of making BFS concrete. The following findings were obtained within the scope of this research on the influence of BFS on the Chloride shielding property of the mortar.

(1) Chloride curing deteriorated slightly through steam curing.

(2) The higher the fineness and addition rate of the BFS, the more effects to the chloride penetration rate in concrete.

5. References

[1] Nippon Slag Association 2015 *Iron and Steel Slag Statistics* 1-7
[2] Takeshi Takeshi Iyoda 2014 *Jpn Ces* **66** 51
[3] Jariyathitipong P, Hosotani K, Fujii T. and Ayano T 2013 *J. Jpn. Soc. Civil Eng. (Materials and Concrete Structures)* **69** 337
[4] Fujii T and Ayano T 2014 *J. Jpn. Soc. Civil Eng. (Materials and Concrete Structures)* **70** 417
[5] Jariyathitipong P, Fujii T. and Ayano T 2014 *Cement Science and Concrete Technology* **67** 427
[6] JIS R 5201 *Physical testing methods of cement.* Translated and published by Japanese Stands Association
[7] JSCE-G571-2003 *Test method for effective diffusion coefficient of chloride ion in concrete by migration.* Japan Society of Civil Engineers
[8] JIS R 5202: 2010 *Methods for chemical analysis of cement*. Translated and published by Japanese Stands Association

IOP Conf. Series: Materials Science and Engineering **840** (2020) 012015 doi:10.1088/1757-899X/840/1/012015

Utilization of Waste Lubricating Oil as a Diesel Engine Fuel

Abhishek Sharma[1,*], Gaurav Gupta[2] and Alok Agrawal[3]

[1]Department of Mechanical Engineering Manipal University Jaipur, Rajasthan, India
[2]School of Mechanical Engineering, VIT University, Vellore, Tamil Nadu, India
[3]Department of Mechanical Engineering, Sagar Institute of Research and Technology-Excellence, Bhopal-462041, India

*Email: drasharma58@gmail.com

Abstract. This research work is focused on the regeneration of used lubricating oils which are simply thrown out to the environment. It is evident from the past research outcome that there is an acute shortage of petroleum oils and regeneration of fuels from used lubricating oils can be potential substitute of fossil fuels. But due to some drawback such as higher viscosity and density it cannot be used as a single fuel. In this regard the distillation of used lubricating oil has been carried out with. Further, this distilled used lubricating oil (DULO) was blended in different proportions with Jatropha biodiesel (JB) which has excess oxygen by about 10%. The aim of the current study is to investigate the performance and emission characteristics of a DI diesel engine run on these blends and comparison with base fuel diesel. The behavior of the engine was evaluated in terms of brake thermal efficiency, brake specific fuel consumption, exhaust gas temperature, carbon monoxide emission, hydrocarbon, nitric oxide emission and smoke opacity. The test result revealed that, the behaviour of the engine run on the JBDULO20 (i.e. 80% JB and 20% DULO) blend was better than other blends considered in the study.

1. Introduction

Energy is a simple but most powerful word and we can say most important thing in this universe. As we know today world is full of needs, full of machines, human being wants there work to be done without any efforts by them so to make that happen we need energy [1]. Energy can be produced by using various available feedstock such as solar, oil, nuclear, coal, fossil fuels etc. for producing electricity and power [2]. Today world's hot topic is renewable energy and how we can reduce greenhouse gas (GHG) emissions [3]. Need of energy is in every field of life heating, lighting, health, food production and storage, education, mineral extraction, industrial production and transportation etc. Today humankind is facing shortage of energy to fulfill the needs of human being because of rapid increase in population. As population increases, we need more energy to fulfill need of people. In 2016, world population was about 7,466,964,280 and in 2017 this number increase to 7,550,262,101 [4]. There is only solution for this is sustainability of energy.

It is essential to increase the usage of renewable energy and should make a decrement in non-renewable energy usage. In world, human beings are using by about 19% [5-6] of renewable energy in the form of biomass, hydropower, solar energy, geothermal energy etc. The research work is continuing to increase the usage of renewable energy and biomass up to 50% so that we can overcome with energy crisis problem. By using these we can make a diversion of usage of energy for many applications such as for vehicles, electricity etc. We can use solar energy, wind energy, biomass energy, geothermal energy for developing electricity.

Lubricating oil is a significant resource and a petroleum base product [7]. The oil is used for many applications which include reducing friction, as shielding against wear in many rotating equipment's. After some time of usage of these lubricating oils, their properties changes which may affects the operation and life of the rotating parts. The undesirable components which accumulate with the oils during their use and which are generally responsible for deterioration are usually of two kinds of which one is extraneous impurities and other is degradation products of oil. The used lubricating oil contains different kinds of impurities in it and is considered one of the harmful liquids which contaminate the environment is simply disposed off.

It is evident from the past research outcome that there is an acute shortage of petroleum fuels [8-10]. The regeneration of fuels from used lubricating oils can be potential substitute of fossil fuels. In this context, the present research work is aimed on the regeneration of used lubricating oils which are simply thrown out to the environment. But due to some drawback such as higher viscosity and density it cannot be used as a single fuel. In this regard the distillation of used lubricating oil has been carried out. Further, this distilled used lubricating oil (DULO) was blended in different proportions (10-40% at regular intervals of 10% on a volume basis) with Jatropha biodiesel (JB) which has high cetane value which will help for enhancing the combustion process. The aim of the current study is to investigate the performance and emission characteristics of a DI diesel engine run on these blends and comparison with base fuel diesel.

2. Material and methods

For the present investigation, Jatropha oil was selected as a raw material to produce Jatropha Biodiesel (JB). In the present study the JB was produced by transesterification process. The transesterification process to produce JB is given in Figure 1. The optimal inputs for the transesterification of Jatropha oil are identified to be 40% methanol and 1.0% NaOH. The maximum ester yield is achieved after 90 min reaction time at 60°C.

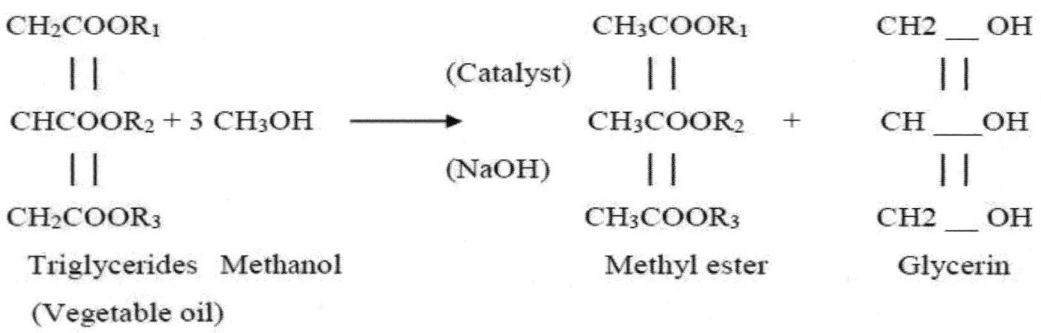

Figure 1. The transesterification process

The purification of used lubricating oil (ULO) was done using atmospheric distillation process. The ULO was heated with a mantle from 160 °C to 180 °C under atmospheric pressure, for an hour. Further, the JB and distilled used lubricating oil (DULO) at low percentages (10-40% at regular intervals of 10% on a volume basis), was blended with the JB at 90 to 60% respectively, to get the fuel blends for the investigation.

3. Experimental setup

The present work was done on a, single cylinder, direct injection diesel engine with a rated power of 4.4 kW at a constant speed of 1500 rpm. The test was started with diesel fuel to record base data and then switched over with JBDULO blends. The designations of the test fuels and their compositions used in this test are provided below in Table 1.

Table 1. Percentage of fuel available in different blends

Blends	JB	DULO
JBDULO10	90%	10%
JBDULO 20	80%	20%
JBDULO 30	70%	30%
JBDULO 40	60%	40%

Cylinder pressure inside the combustion chamber was measured by using piezoelectric pressure transducer connected to piezocharge amplifier. All the sensors were attached with a control panel, which communicate all the results to the computer. All the inputs such as engine brake power, air and fuel consumption were recorded by data acquisition system, stored in computer and displayed in the monitor. The schematic diagram of the experimental setup and specification are given in figure 2 and Table 2 respectively.

Table 2. Technical Specifications of Test Engine

Engine Manufacturer/Model	**Kirloskar/** TAF 1
Bore and Stroke	87.5 mm and 110 mm
Compression ratio	17.5:1
Number of cylinder	1
Piston Type	Bowl-in-piston
Maximum power output	4.4 kW
Maximum speed	1500 rpm (Constant)
Type of fuel injection	Pump-line-nozzle injection system
Nozzle type	Multi hole (3 hole)
Start of fuel injection	23 °CA bTDC
Fuel injection pressure	200 bar
Type of Cooling	Air cooled with radial fan

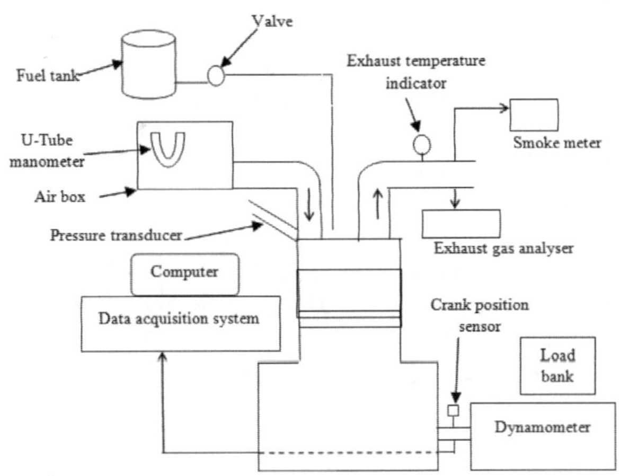

Figure 2. Schematic representation of test setup

4. Results and discussions

4.1. Performance analysis

The important performance terms of a diesel engine are discussed in following section.

4.1.1. Brake thermal efficiency. The variation of brake thermal efficiency (BTE) with respect to brake power for all the test fuels is shown in figure 3.

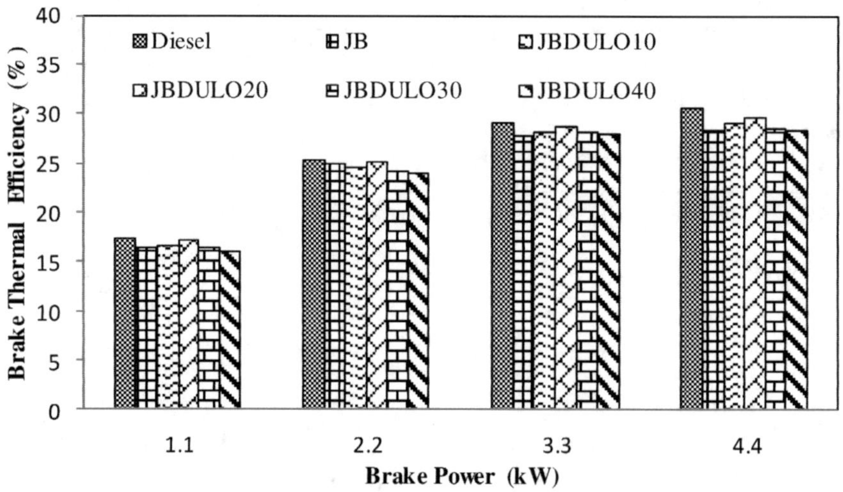

Figure 3. Variation of brake thermal efficiency with brake power

The brake thermal efficiency is the ratio of power output or work done by the engine with respect to heat supplied to it [11]. It is also called as fuel conversion efficiency. The thermal efficiency of the engine increases with increase in brake power for all the fuel operations [12-13]. This is because, while increasing the load, both the brake power (BP) and fuel flow rate increases and reduction in heat loss occur at higher loads. It is highly depending on some parameters like calorific value, fuel consumption. The BTE values for diesel, JB, JBDULO10, JBDULO20, JBDULO30, and JBDULO40 are 30.6%, 28.4%, 29.13%, 29.4%, 28.42% and 28.12% respectively at full load condition. Figure 3 clearly says that the higher brake thermal efficiency for diesel is 30.6% and JBDULO20 is 29.4%, which is 7% maximum brake thermal efficiency when compared with the neat Jatropha biodiesel. The BTE of the engine depends primarily on the calorific value of the fuel used [14]. The decrease of DULO concentration BTE also increases due to high level oxygen content in JB; it increases the better atomization, maximum heat release rate, higher pressure and better performance [15].

4.1.2. Brake specific energy consumption. The BSFC is the mass of fuel consumed per unit power output in a diesel engine. It is correlated with viscosity, CV, CN, heating content and specific gravity [16-17]. The figure 4 represents the variation of BSEC with brake power in the diesel engine with four different DULO percentages test fuel are, JB+DULO10%, JB+DULO20%, JB+DULO30%, and JB+DULO40% and compared with diesel.

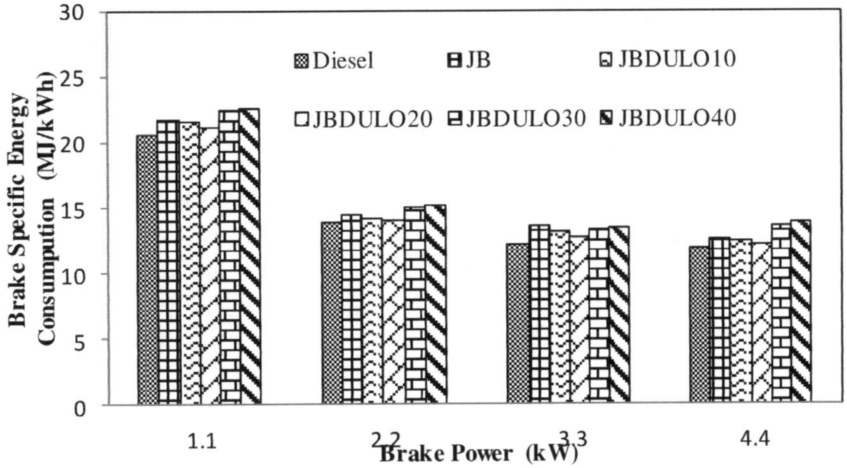

Figure 4. Variation of brake specific energy consumption with brake power

The BSEC changes from 0.391 kg/kW h at low load to 0.263 at high load condition for normal diesel fuel, and its changes from 0.413 kg/kW h at low load condition to 0.266 at high load condition for JBDULO10, 0.423 kg/kW h at low load to 0.278 g/kWh at high load condition for JBDULO20, 0.441 kg/kW h at low load to 0.291 high load condition for JBDULO30%, 0.541 kg/kW h at low load to 0.391 for JBDULO40%, respectively. From the figure 4 it says that as the DULO concentration increases, the BSFC also increased from zero to full load condition. This is due to the low heating value of DULO fuel and low cetane number, to main the same power output during power stroke more fuel is required [18]. The CV, latent heat of evaporation, flash point is plays major part in combustion, performance characteristics [19].

4.1.3. Exhaust gas temperature. Figure 5 shows the deviation of exhaust gas temperature (EGT) with respect to brake power for different fuels used in this test. The EGT increases with increase in brake power for all the test fuels. The analysis of EGT is important as it describes about those heat which are not utilized for producing power [20]. The highest EGT value with respect to all the powers was recorded for the blend which has higher percentage of DULO. This result is expected due to present of moisture content in DULO. But up to 20% DULO in blend, the value of EGT of JBDULO20 blend is close to diesel reading.

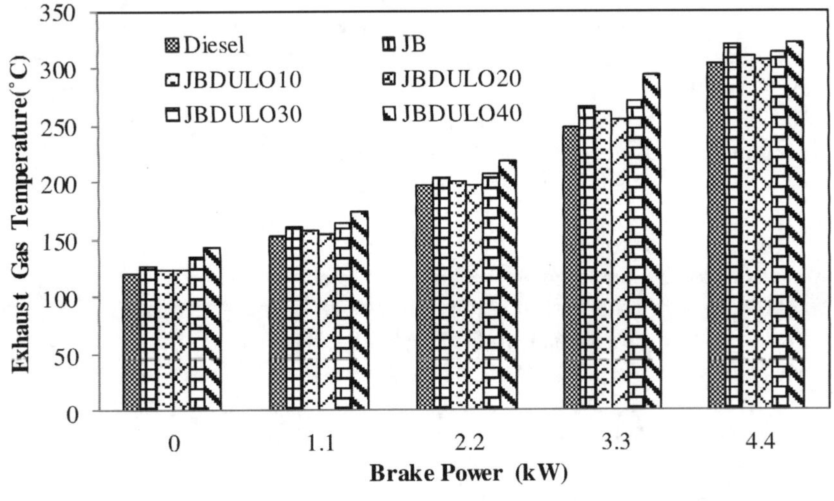

Figure 5. Variation of exhaust gas temperature with brake power

This is due to the availability of oxygen content in the JB which enhances the combustion process which resulted in lower EGT. At a brake power of 4.4 kW (maximum load condition), the EGT of JBDULO40 is 8% higher than neat diesel fuel respectively.

4.2 Emission analysis

4.2.1. Carbon monoxide emission. The chemical reaction takes place between oxygen molecule and carbon molecule during the combustion process the intermediate combustion product is called as CO [21]. It is produced due to insufficient air flow, rich air fuel mixture, etc. The figure 6 shows the development of CO emissions at different brake power for different test fuels and compared with diesel.

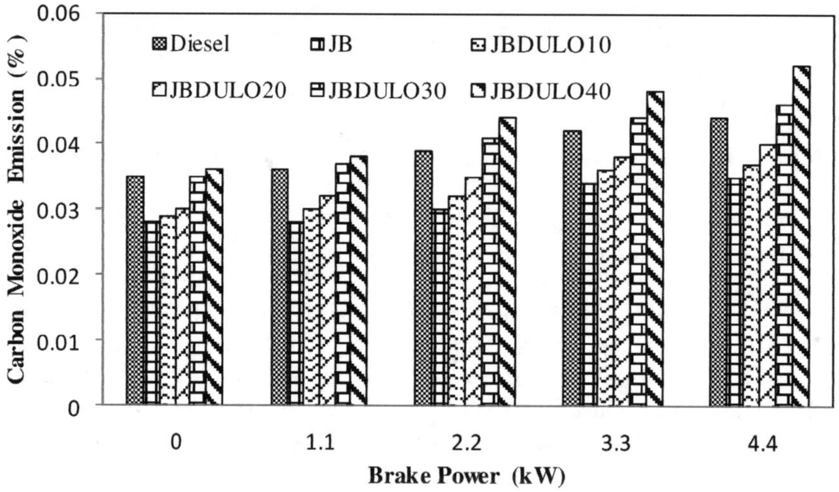

Figure 6. Variation of CO emission with brake power

The value of CO emission changes from 4.96 g/kW h at low load to 3.61 g/kWh at high load condition for normal diesel fuel, and its changes from 5.12 g/kW h at low load condition to 3.85 at high load condition for JBDULO10, 5.49 g/kW h at low load to 3.92 g/kWh at high load condition for JBDULO20%, 5.84 g/kW h at low load to 4.16 at high load condition for JBDULO20, 6.84 g/kW h at low load to 5.16 at high load condition JBDULO40 respectively. The lower CO emissions are liberated at 50% load condition and the values are 1.68 g/kW h for diesel. From the above graph clearly understood that CO emissions increased simultaneously low load condition by 3.22%, 10.68 %, 17.74% and 21.23 to final load condition by 6.6%, 8.5%, 15.23% and 18.2% compared with diesel fuel. The biodiesel has higher lower heat of evaporation decreased the heat transfer, which cause the low cooling effect on the combustion chamber [22]. The overall efficiency is increased, and CO emission is decreased. The reason is rapid addition of DULO in the Jatropha biodiesel blends reduces the overall cetane number when compared with diesel fuel [23-24].

4.2.2. Unburned hydrocarbon emission. The UHC emission shows the quality of the combustion process and amount of oxygen percentages available in the entire conversion of fuel into water vapour and carbon dioxide. The fuel properties, operation condition, incomplete combustion (rich or lean fuel mixture), heat losses around the cylinder, and flame quenching are the important parameters for the formation of UHC emissions [25-26]. The figure 7 shows the formation of UHC at different brake power for different test fuels used in this investigation. The value of UHC emission changes from 2.21 g/kWh at low load to 0.91 g/kWh at high load condition for normal diesel fuel, and its changes from 2.38 g/kW h at low load condition to 0.96 g/kWh at high load condition for JBDULO10, 2.54 g/kW h at low load to 1.05 g/kWh at high load condition for JBDULO20, 2.96 g/kW h at low load to 1.16

g/kWh at high load condition for JBDULO30 and 3.96 g/kW h at low load to 2.16 g/kWh JBDULO40 respectively. From the figure 7, it is observed that UHC increased simultaneously from low load condition by 7.6%, 14.93 % to final load condition by 5.4%, 15.38%, 20.87% compared with diesel fuel. Basically, the UHC emissions are maximum at low load condition than full load condition due to air fuel mixture, minimum cylinder surrounding temperature.

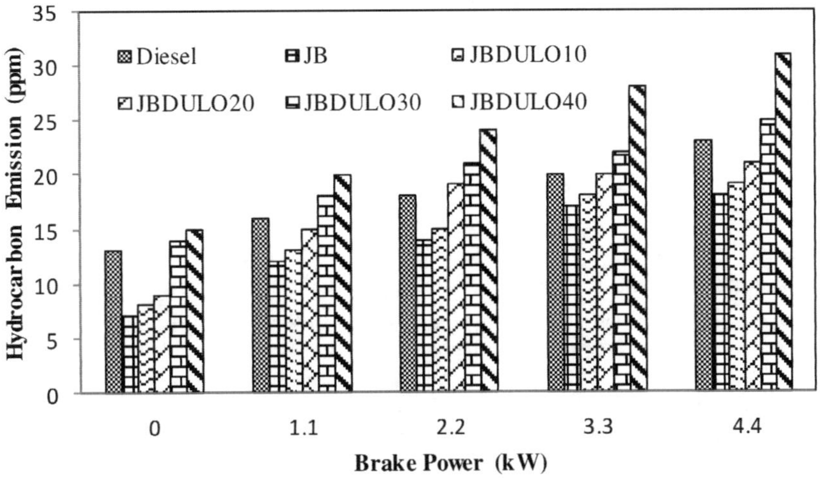

Figure 7. Variation of HC emission with brake power

The addition of DULO percentages with Jatropha biodiesel blend the UHC also increased entire load condition, this is because of low cetane number of DULO reduced their self-ignition properties, higher latent heat of vaporization, more cooling effect are the main reason for incomplete combustion and higher UHC emission [27-28].

4.2.3. Nitric oxide emission. Oxides of nitrogen emission levels are produced when the inert nitrogen molecules react with oxygen molecule at maximum temperature above 1700 K and pressure. Basically, these liberated NOx particles are splits into some wings such as nitric oxide (NO) contain 90%, NO_2 and N_2O at 5% and N_2O_3 and N_2O_5 [29]. Figure 8 compares the variation of NO_x emissions with engine load for JBDULO blends.

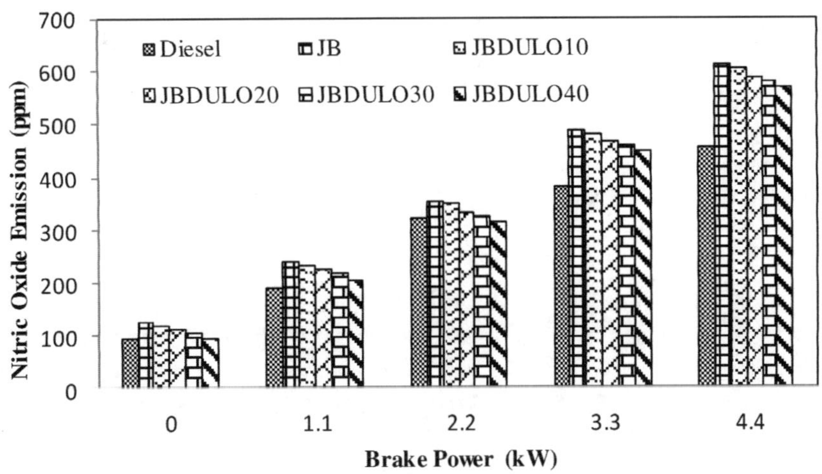

Figure 8. Variation of NO emission with brake power

The value oxides of nitrogen emission changes from 91 ppm at low load to 460 ppm at high load condition for normal diesel fuel, and its changes from 120 ppm at low load condition to 583 ppm at high load condition for JBDULO10, 117 ppm at low load to 570 ppm at high load condition for JBDULO20, 115 ppm at low load to 561 ppm at high load condition for JBDULO20, respectively. The NOx emissions are reduced by 17.31% at JBDULO20, at full load condition compared with JB. By increases of DULO concentration with JB brings down the oxides of nitrogen emissions upto JBDULO20 compare with diesel. This is due to the higher cooling effect liberated by latent heat of vaporization of DULO; it is highly useful to reduce development of oxides of nitrogen emissions [30]. Further increases the DULO concentration in JB causes high cylinder pressure, cylinder temperature increases the NOx formation.

4.2.4 Soot concentration. The results of smoke opacity for different test fuels with respect to brake power is shown in Fig. 9. It is the fact that smoke formation depends upon the load of the engine. Hence, smoke density increases with an increase in load of the engine. The minimum and maximum smoke densities produced during the test for diesel, JB and JBDULO blends were in the range of 4% and 65% respectively. At full load condition, the maximum increase of smoke density was observed for diesel. It is perceived from the figure that the smoke density value increases with the increase in the percentage of DULO in the JB-DULO blends, which is higher than the smoke density of pure JB [31]. The smoke density of diesel and JB is recorded as 65% and 54.5% at rated load. In case of DULO10, DULO20, DULO30 and DULO40 blends the values are 55.2%, 57.4%, 60.4% and 64.9% at full load. The increase in smoke emission is caused by poor atomization of the DULO molecule present in the blend. The factors which cause higher amount of smoke emission are the higher viscosity, bulk fuel molecules, low volatility and presence of aromatic of waste plastic oil constituted in reduced atomization of fuel.

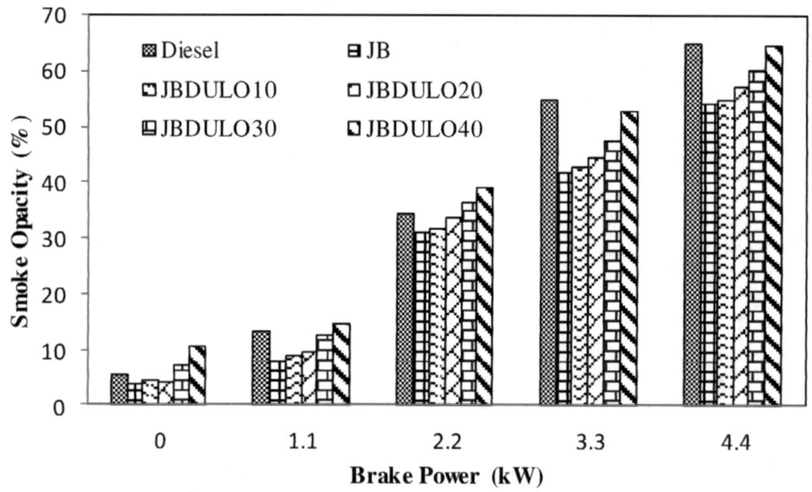

Figure 9. Variation of smoke density with brake power

5.Conclusion

The main aim of this study is to improve the performance of a diesel engine when distilled used lubricant oil (DULO) is used as partial replacement of diesel/biodiesel fuel. In this regard behaviour of a single cylinder, diesel engine fuelled with JBDULO blend as a fuel were tested and results obtained were equated with base line reading of diesel. The conclusions of this study are given in this section.

- The BTE value of the JBDULO20 blend was 29.1% while for diesel was 30.5% at full load.
- The energy losses in the form of EGT were higher for JBDULO20 blend in comparison with diesel operation.

- The value of BSEC for diesel was 11.7 MJ/kWh at full load. The increase in 4.3% of BSEC was recorded for JBDULO20, than that of diesel reading.
- The CO, HC and smoke emissions were reduced by 8.7%, 7.3% and 12.4% respectively for JBDULO20 blend, compared to diesel.
- The NO emission was recorded extra by about 15.3% for JBDULO20, in comparison with diesel.
- By considering the above results, it can be summarised that the JBDULO20 blend revealed better engine results compared to those of other JBDULO20 blends. Overall it can be suggested that this blend can be used as a replacement of diesel fuel in a diesel engine.

6. Reference

[1] Kramer DM, Evans JR. The importance of energy balance in improving photosynthetic productivity. Plant Physiology. 2011 Jan 1;155(1):70-8.

[2] Falk G, Herrmann F, Schmid GB. Energy forms or energy carriers?. American Journal of Physics. 1983 Dec;51(12):1074-7.

[3] Weisser D. A guide to life-cycle greenhouse gas (GHG) emissions from electric supply technologies. Energy. 2007 Sep 1;32(9):1543-59.

[4] Alvaredo F, Chancel L, Piketty T, Saez E, Zucman G, editors. World inequality report 2018. Belknap Press; 2018.

[5] Li X. Diversification and localization of energy systems for sustainable development and energy security. Energy policy. 2005 Nov 1; 33(17):2237-43.

[6] Hadjipaschalis I, Poullikkas A, Efthimiou V. Overview of current and future energy storage technologies for electric power applications. Renewable and sustainable energy reviews. 2009 Aug 1; 13(6-7):1513-22.

[7] L. Andrews, "Compendium of Recycling and Destruction Technologies for Waste Oils," United Nations Environment Programme, 2008.

[8] Sharma A, Murugan S. Experimental evaluation of combustion parameters of a DI diesel engine operating with biodiesel blend at varying injection timings. In Proceedings of the first international conference on recent advances in bioenergy research 2016 (pp. 169-177). Springer, New Delhi.

[9] Sharma A, Murugan S. Effect of nozzle opening pressure on the behaviour of a diesel engine running with non-petroleum fuel. Energy. 2017 May 15; 127:236-46.

[10] Sharma A, Murugan S. Effect of blending waste tyre derived fuel on oxidation stability of biodiesel and performance and emission studies of a diesel engine. Applied Thermal Engineering. 2017 May 25; 118:365-74.

[11] Sellnau M, Foster M, Moore W, Sinnamon J, Hoyer K, Klemm W. Pathway to 50% Brake Thermal Efficiency Using Gasoline Direct Injection Compression Ignition. SAE Technical Paper; 2019 Apr 2.

[12] Kinnal N, Sujaykumar G, D'costa SW, Girishkumar GS. Investigation on Performance of Diesel Engine by Using Waste Chicken Fat Biodiesel. InIOP Conference Series: Materials Science and Engineering 2018 Jun (Vol. 376, No. 1, p. 012012). IOP Publishing.

[13] Mahalingam A, Munuswamy DB, Devarajan Y, Radhakrishnan S. Emission and performance analysis on the effect of exhaust gas recirculation in alcohol-biodiesel aspirated research diesel engine. Environmental Science and Pollution Research. 2018 May 1;25(13):12641-7.

[14] Devarajan Y, Munuswamy DB, Mahalingam A. Influence of nano-additive on performance and emission characteristics of a diesel engine running on neat neem oil biodiesel. Environmental Science and Pollution Research. 2018 Sep 1;25(26):26167-72.

[15] Ray SK, Prakash O. Biodiesel Extracted from Waste Vegetable Oil as an Alternative Fuel for Diesel Engine: Performance. Renewable Energy and Its Innovative Technologies: Proceedings of ICEMIT 2017. 2018; 1:219.

[16] Wasiu S, Aziz RA, Megat P. Brake Specific Energy Consumption (BSEC) and Emission Characteristics of the Direct Injection Spark Ignition Engine Fuelled by Hydrogen Enriched

Compressed Natural Gas at Various Air-Fuel Ratios. International Journal of Applied Engineering Research. 2018;13(1):677-83.
[17] Nabi MN, Rasul MG. Influence of second-generation biodiesel on engine performance, emissions, energy and exergy parameters. Energy conversion and management. 2018 Aug 1; 169:326-33.
[18] Noce T, da Silva RR, Morais R, Sales LC, de Morais Hanriot S, Sodré JR. Energy factors for flexible fuel engines and vehicles operating with gasoline-ethanol blends. Transportation Research Part D: Transport and Environment. 2018 Dec 1; 65:368-74.
[19] Akula N, Rajagopal K, Murthy KD. Experimental Investigation of Bio-Gas Fuel used as Renewable Energy Source for SI Engine.
[20] Heskett S, Hunter WO, Zhirov V, inventors; H & H Inventions & Enterprises Inc, assignee. Exhaust gas purification method and system. United States patent application US 10/245,546. 2019 Apr 2.
[21] Mahalingam AR, Munuswamy DB, Devarajan Y, Radhakrishnan S. Investigation on the emission reduction technique in acetone-biodiesel aspirated diesel engine. Journal of Oil Palm Research. 2018 Jun 1;30(2):345-9.
[22] Rathinam S, Justin Abraham Baby S, Devarajan Y. Influence of water on exhaust emissions on unmodified diesel engine propelled with biodiesel. Energy Sources, Part A: Recovery, Utilization, and Environmental Effects. 2018 Nov 2;40(21):2511-7.
[23] Arul Gnana Dhas A, Devarajan Y, Nagappan B. Analysis of emission reduction in ethyne–biodiesel-aspirated diesel engine. International Journal of Green Energy. 2018 May 28;15(7):436-40.
[24] Devarajan Y, Mahalingam A, Munuswamy DB, Nagappan B. Emission and combustion profile study of unmodified research engine propelled with neat biofuels. Environmental Science and Pollution Research. 2018 Jul 1;25(20):19643-56.
[25] Mahalingam A, Munuswamy DB, Devarajan Y, Radhakrishnan S. Emission and performance analysis on the effect of exhaust gas recirculation in alcohol-biodiesel aspirated research diesel engine. Environmental Science and Pollution Research. 2018 May 1;25(13):12641-7.
[26] Sivaramakrishnan K. Investigation on performance and emission characteristics of a variable compression multi fuel engine fuelled with Karanja biodiesel–diesel blend. Egyptian Journal of Petroleum. 2018 Jun 1;27(2):177-86.
[27] Devarajan Y, Mahalingam A, Munuswamy DB, Nagappan B. Emission and combustion profile study of unmodified research engine propelled with neat biofuels. Environmental Science and Pollution Research. 2018 Jul 1;25(20):19643-56.
[28] Bragadeshwaran A, Kasianantham N, Kaisan MU, Reddy DM, Aravind KM, Paul N, Ali IM, Jose A, Chungath T. Influence of injection timing and exhaust gas recirculation (EGR) rate on lemon peel oil–fuelled CI engine. Environmental Science and Pollution Research. 2019 May 28:1-5.
[29] Bragadeshwaran A, Kasianantham N, Kaisan MU, Reddy DM, Aravind KM, Paul N, Ali IM, Jose A, Chungath T. Influence of injection timing and exhaust gas recirculation (EGR) rate on lemon peel oil–fuelled CI engine. Environmental Science and Pollution Research. 2019 May 28:1-5.
[30] Bragadeshwaran A, Kasianantham N, Kaisan MU, Reddy DM, Aravind KM, Paul N, Ali IM, Jose A, Chungath T. Influence of injection timing and exhaust gas recirculation (EGR) rate on lemon peel oil–fuelled CI engine. Environmental Science and Pollution Research. 2019 May 28:1-5.
[31] Radhakrishnan S, Munuswamy DB, Devarajan Y, Mahalingam A. Effect of nanoparticle on emission and performance characteristics of a diesel engine fueled with cashew nut shell biodiesel. Energy Sources, Part A: Recovery, Utilization, and Environmental Effects. 2018 Oct 18;40(20):2485-93.

IOP Conf. Series: Materials Science and Engineering **840** (2020) 012016 doi:10.1088/1757-899X/840/1/012016

Utilisation of Eggshell Powder (ESP) as Partial Replacement of Cement Incorporating Superplasticizer

Gunalaan Vasudevan[1], Seah Chan Wei[2]

[12]Department of Construction Management, Faculty of Built Environment, Tunku Abdul Rahman University College Malaysia

Email: gunalaanv@tarc.edu.my

Abstract. This research showed the results of experiments evaluating the use of eggshell powder from egg production industry as partial replacement for ordinary Portland cement. Research on the reuse of waste materials in the concrete industry has been quite intensive in the past decade. The objective of this research is to identify the performance of dried eggshell powder as a partial cement replacement in the production of concrete. Eggshell powder of various amounts, namely 5%, 10%, 15% and 20% by volume, was added as a replacement for ordinary Portland cement. The results showed that eggshell concrete greatly improved the compressive and flexural strength of concrete. The rate of water absorption of eggshell concrete was reduced as eggshell powder filled up the existing voids, making it more impermeable. However, the compressive strength of the eggshell concrete decreases gradually when the amount of eggshell powder increased. It can be concluded that the optimum percentage of dried eggshell powder as a partial cement replacement is 15%. In this direction, an experimental investigation of ultrasonic pulse velocity, rebound hammer concrete test, compressive strength, flexural strength and FTIR spectra and TGA analysis was undertaken to use eggshell powder and admixtures as partial replacement for cement in concrete.

1. Introduction

Concrete is one of the most significant natural resource products. The use of concrete in the modern era was the foundation of the nation's construction industry. The local construction industry growth requires greater concrete quantity, which indirectly increases the need for greater supply of natural resources for concrete production. A change in cement can be obtained on the market and each type is used under persuaded illness owing to its unique characteristics such as cement colour and different way of arrangement. In latest years, there has been a general trend towards reducing the use of natural resources and the reuse of waste products. This experiment is to examine the performance of chicken eggshell blended into eggshell powder as potential replacement and chicken eggshell waste used as a limestone filler in Portland cement mortar [1]. A handling technique has been created to generate big amounts of eggshells. Limestone is introduced as replacement of cement. Measurements of compressive test and flexural strength are performed. Eggshell consists several growing layer of calcium carbonates. The innermost layer-maxillary 3 layer develops on the outermost egg membrane and produces the core on which the thickest portion of the eggshell at the palisade layer. The quality of eggshell waste is strongly affected by the sunlight exposure, raw water and severe conditions. Before using with concrete or mortar, the fine-grained powder with the appropriate proportion that is sieved to the required size [2].

IOP Conf. Series: Materials Science and Engineering **840** (2020) 012016 doi:10.1088/1757-899X/840/1/012016

2. Problem statement

Eggshell is referred to as waste products or by-products from hatcheries and food sectors. Majority of the eggshell was disposed of in the landfill without any pre-treatment since it is recognized as useless traditionally. In Malaysia, water disposal and land filling are one of the severe problems. It is observed that Malaysia has very high consumption of chicken and eggs among all the Asian countries. The number of landfills and waste disposal increase significantly from years to years. If the waste disposal did not dispose in a correct way, it will lead many negative impacts in terms of environmental pollution and water pollution [3]. Compared to European nations, the use of food waste such as eggshells in Malaysia is very limited. All the landfill had waste related to aesthetic problem, contamination, landfill gas or odour problems. Almost all the waste is disposed using landfill method and majority of the sites have poor management. Discovered an alarming increase in the dumping site in Malaysia, which encouraged the Malaysian government to open up more dumping sites resulting in soil scarcity. Land scarcity has dramatically increased the price of landfill. Identify material from waste and use it in concrete production in construction industry could be a wide idea and bringing a lot of benefits as Portland cement is very costly, energy intensive and produce great amounts of carbon emission [4].

3. Literature review

3.1. Physical properties of eggshell powder

Egg Shell Powder (ESP) can be used as an industrial lime supplement on expansive clay land and it has also been reported that the mixture can be used where elevated subgrade output is not essential. Conducted the inquiry into eggshell waste. It can be used as an alternative raw material in the manufacture of wall tile goods based on the presence of $CaCO_3$ in eggshells. They also found that eggshells can be used as an exceptional choice for techniques of reuse of materials and waste recycling [5]. Eggshell powder concrete properties as a substitute for cement, this research discusses studies on the use of concrete poultry waste by developing concrete that incorporates eggshell powder (ESP). By exchanging 5-15% of ESP for cement, different ESP concretes were created [6]. The findings showed that ESP can be used effectively in concrete manufacturing as a partial substitute of cement. The information provided contains the growth of strength and characteristics of transportation. Regarding the outcomes, the yields were greater than the control material at 5% ESP substitute and suggest that 5% ESP is an optimal material for peak strength. Five mixes were used in this research to explore the characteristics of ESP concretes.

3.2. Chemical properties of eggshell powder

Experimentally examined the eggshell as a temporary substitute for concrete cement. This document records the outcomes of studies assessing the use of eggshell powder from the egg sector as a partial substitute in cement mortar for ordinary Portland cement. It was determined the chemical composition of the eggshell powder and the concrete mix compressive strength. Mixed concrete ratio 1:2:4, in which cement is partially replaced with eggshell powder as 5%, 10%, 15%, 20%, 25% by cement weight. The compressive strength was determined at 28 days of curing age. Compressive strength dropped sharply beyond 5 percent eggshell-powder replacement. Saw dust ash, fly ash and micro silica are used as admixtures in order to increase the power of the concrete mixture with 5% eggshell powder as a partial substitute for cement. In this path, an experimental study was carried out on compressive strength [7], divided tensile strength and flexural strength to use eggshell powder and admixtures as a temporary substituted for concrete cement.

3.3. Use of eggshell powder on construction materials

Several scientists have tried to use waste products to enhance concrete's mechanical characteristics and decrease waste disposal issues. The major waste products used to date are fly ash, silica fumes, quartz sand, eggshell powder [3]. Eggshell disposal is said to add to environmental pollution, producing undesirable odors that cause discomfort and effect on human well-being. Calcium carbonate ($CaCO_3$) is the primary component in eggshells. The shell itself is said to consist of 95% $CaCO_3$ while the remaining components consist of magnesium, aluminum, phosphorous, salt, sulfur, mercury, iron, salt, ironic acid and silica acid. Eggshells can function as a healthy bio-sorbent because they have a cellulosic framework with amino acids [2]. Using eggshell will help reduce alkali-silica and sulphate expansions. It has been shown that eggshells have great durability and a washable surface. On paint movie, Eggshell can withstand mould and mildew. The use of eggshells has economic advantages and the ability to satisfy both rigorous efficiency and aesthetic requirements [8].

4. Result and discussion

4.1. Compressive strength test

Based on Figure 1, show that the compressive strength started to increase at 5% of eggshell powder replacement compared to the control mixed design. The compressive strength is increase after added into the eggshell powder and the highest value is 26.6 MPa at 28th days of curing and the lowest value is 22 MPa at 7th days curing. At the 14th days of curing, the compressive strength is higher than 7th days of curing. The reason is because as the curing process is longer, it will increase the workability and durability of concrete. Observed that the strength increased with the curing age and the percentage of eggshell replacement. The strength is stronger than the control mix design because the eggshell powder is added into the cement. It is interesting to note that eggshell concrete exhibits high early compressive strength compared with the control specimen at the early stage of curing. Partial replacement of cement with eggshell powder improved the reaction between silica from the cement and calcium oxide from the eggshell powder, a by-product of the cement hydration process in the continuous presence of moisture. In addition, the eggshell powder also acted as a filler by filling up the existing voids, which made the internal structure of the concrete more packed and led to the development of higher compressive strength. However, further increase of ESP will decrease compressive strength at all curing ages.

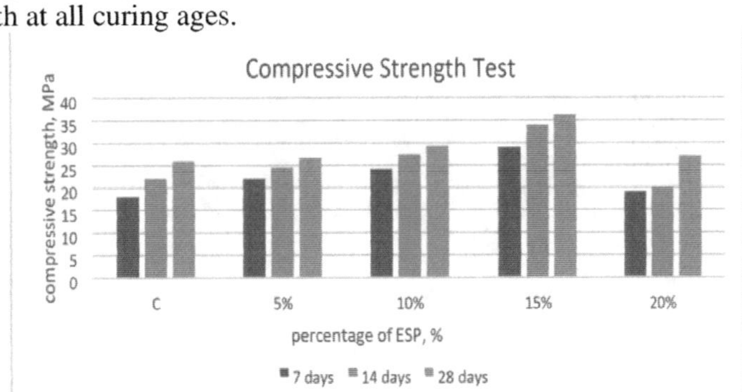

Figure 1. Compressive strength test

4.2. Ultrasonic pulse velocity (UPV)

The results UPV test of high performance concrete made from based on the Figure 2 above, at 7th days of curing, the highest value of ultrasonic pulse velocity is showed at 15% of eggshell powder replacement which is stated as 45.4m/s-1. At 14th days, the value is increase compared to 7th days of curing, and figure above was observed that 15% replacement of eggshell powder and at 14th which given data for direct test is 43.4 m/s-1. 14 days of curing has higher value than 7th days of curing but lower than 28th days of curing. At the age of 28th days curing, 15% replacement of eggshell powder is achieving the highest ultrasonic pulse velocity value which is 45.9 m/s-1 if compared to other percentage replacement of eggshell powder on 28th days. There is a huge change of value from 0% to 15% of increase and a value drop at 20% of eggshell powder replacement.

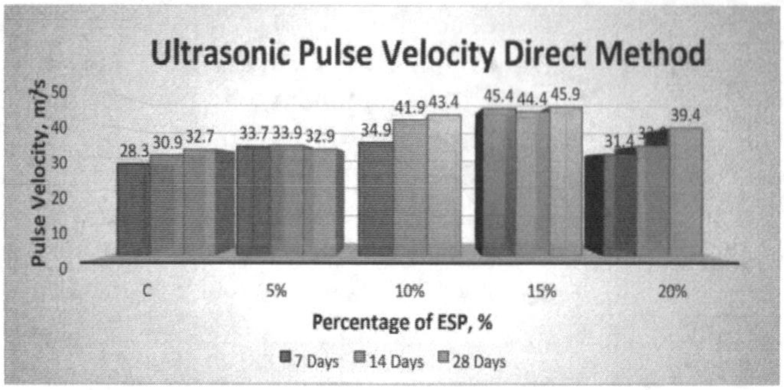

Figure 2. Results of UPV

4.3. Flexural strength test

Based on the Figure 3, the value is started to increase after the eggshell powder is added into the concrete. At the age of 7 days curing, the highest value of flexural strength is 2.7 MPa. It can see that the flexural strength is started to increase compared to control mix design. At the age of 14th days of curing, the value is flexural strength is increase as the curing process is longer. The lowest value of flexural strength is 2.2 MPa and the highest value is 2.9 MPa. At the age of 28 d, 15% exhibits the highest flexural strength, which is higher than that of control mixed design. Thus, the eggshell powder seems to have greatly improved the flexural strength of concrete with a partial cement replacement of up to 15%. However, the value will drop at 20% of eggshell powder replacement but still has higher value compared to control mix design. In addition, calcium hydroxide contributes slightly to the strength and impermeability of the concrete because it reduces the total pore volume by converting some of the liquid water into solid.

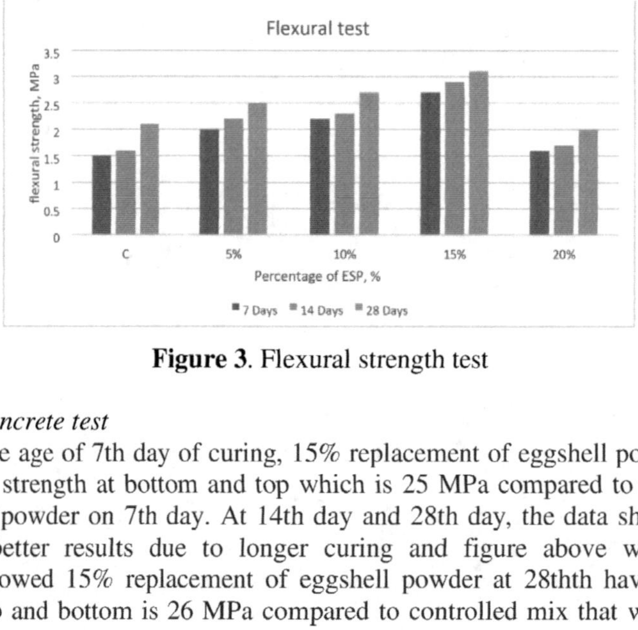

Figure 3. Flexural strength test

4.4. Rebound hammer concrete test

The Figure 4, show at the age of 7th day of curing, 15% replacement of eggshell powder is achieving the highest compressive strength at bottom and top which is 25 MPa compared to other percentages replacement of eggshell powder on 7th day. At 14th day and 28th day, the data showed that the 7th day have higher and better results due to longer curing and figure above was observed that compressive strength showed 15% replacement of eggshell powder at 28thth have the best results which given data for top and bottom is 26 MPa compared to controlled mix that without add in any eggshell powder.

IOP Conf. Series: Materials Science and Engineering **840** (2020) 012016 doi:10.1088/1757-899X/840/1/012016

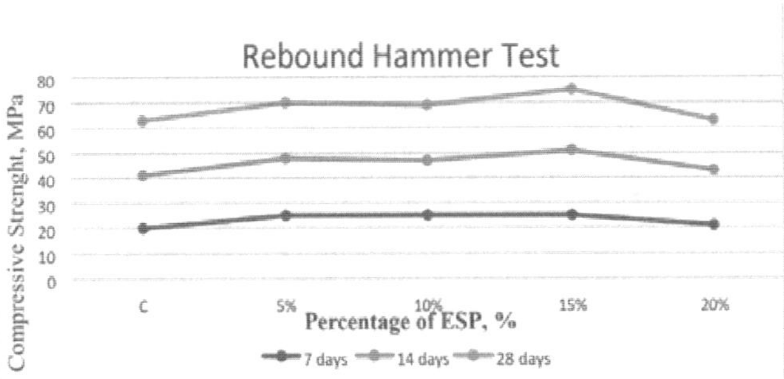

Figure 4. Rebound hammer test

4.5. Microstructures of eggshell

The apparent morphologies of the eggshell examined by SEM are shown in Figure 5. The control mix concrete comprises irregular shape of particles and finely modulated, which was showed in Figure 5(a). Meanwhile, Figure 5(b) showed the microstructure of eggshell. After mix with concrete, the structure of eggshell has been changed from irregular shape to interconnected like skeleton structure. Besides that, the catalyst resulted in strong increase in active sites, which supported by SEM images [9].

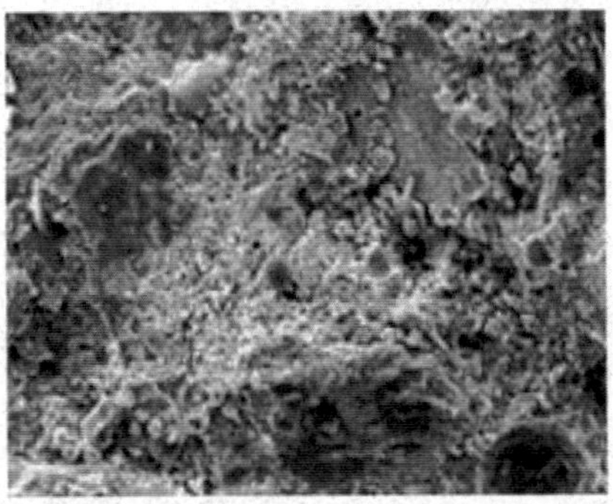

Figure 5(a). SEM images of control mix

Figure 5(b). SEM images of Eggshell

4.6. FTIR spectra and TGA analysis

IR spectra of both eggshell were determined by fourier transform infrared (FTIR) spectrometry. Based on the results from Figure 6(a), IR spectra for eggshell showed wide stretching peaks at 1395.7 cm-1 attributing to C-H bonding. Sharp and intense peak observed at 871 cm-1represent C-O stretching and bending modes of $CaCO_3$. Meanwhile, the sharp band at 710 cm-1 represent Ca-O bonds. Figure 6(b) shows the TGA curved of eggshell. Result from TGA showed that there were two stages of weight losses. For the first stage, minor weight loss occurred at temperature below 300 °C - 500°C attributed to an adsorption of water molecules to the surface of the material and loss of organic compound. [10]

IOP Conf. Series: Materials Science and Engineering **840** (2020) 012016 doi:10.1088/1757-899X/840/1/012016

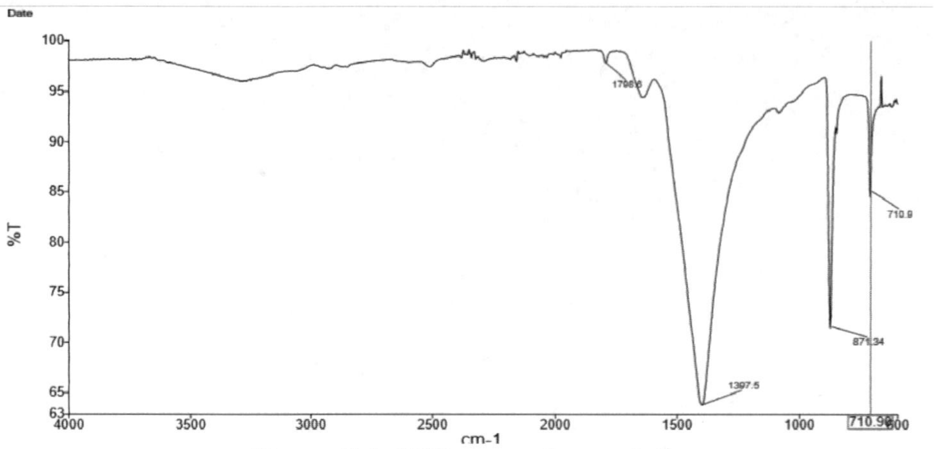

Figure 6(a). FTIR spectra for eggshell

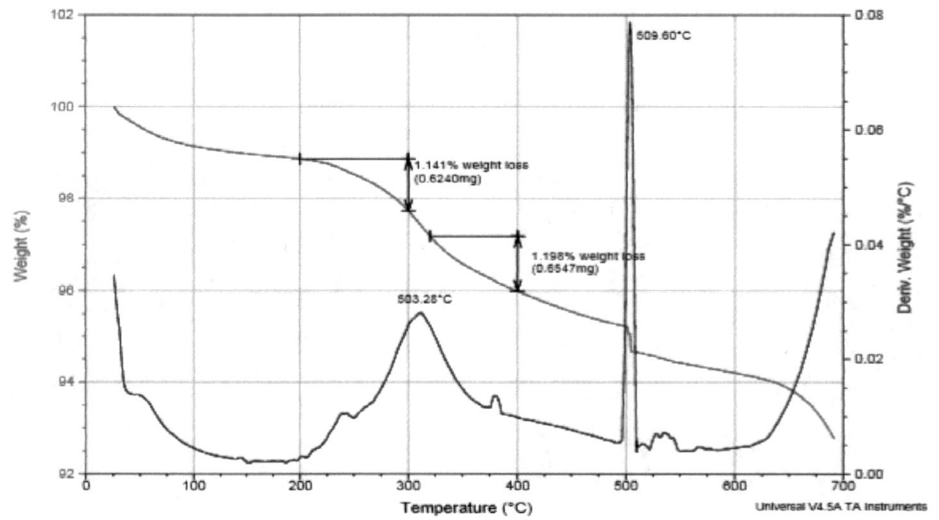

Figure 6(b). TGA analysis for eggshell

5. Conclusion and recommendation

The overall objective of the work was to investigate the feasibility of utilisation of eggshell powder as an additive material to concrete mix. This study included the preparation of concrete mixes containing and the evaluation of eggshell powder concrete properties in fresh and hardened states. The studying properties involved mix workability, compressive strength and density. According to experimental results, the usage of eggshell powder in concrete mixes as an alternative of disposal for eggshell powder is possible. The current type of eggshell powder accumulated in dumping sites and the expected future type were used in making concrete mixes. The influence of both types on concrete properties was studied. In all cases there was an optimum quantity of eggshell powder which can be used without introducing any change in mix preparation and acceptable properties were still be produced. The study showed that the eggshell powder can be used in interlock brick production without changing the normal industrial process. Recommendation for further research The following recommendations are proposed for further research and study in order to from a complete picture of using eggshell powder in concrete mixes:

1) Further investigation is needed on the different superplasticizer used together with eggshell powder in the concrete mix design.

2) Further study is required for eggshell powder mix with other by-products in the concrete mix design.

3) Further investigation of the higher percentage of replacement of the cement by the eggshell powder in concrete mix design.

4) Longer curing time is required in order to understand the behaviour of the concrete, and it might give the different result for eggshell powder concrete.

6. Reference

[1] Adogla, F., YALLEY, P. and ARKOH, K., 2016. Improving compressed laterite bricks using powdered eggshells. Int. J. Eng. Sci, 5(4), pp.65-70.

[2] Heikal, M., 2016. Characteristics, textural properties and fire resistance of cement pastes containing Fe_2O_3 nano-particles. Journal of Thermal Analysis and Calorimetry, 126(3), pp.1077-1087

[3] Al-Jumaily, I.A.S., Naji, N. and Kareem, Q., 2015. An overview on the Influence of Pozzolanic Materials on Properties of Concrete. International Journal of Enhanced Research in Science Technology & Engineering, 4(3), pp.81-92.

[4] McLellan, B.C., Williams, R.P., Lay, J., Van Riessen, A. and Corder, G.D., 2011. Costs and carbon emissions for geopolymer pastes in comparison to ordinary portland cement. Journal of cleaner production, 19(9-10), pp.1080-1090.

[5] Najjar, M.F., Soliman, A.M. and Nehdi, M.L., 2016. Two-stage concrete made with single, binary and ternary binders. Materials and Structures, 49(1-2), pp.317-327

[6] Chen, X., Wu, S. and Zhou, J., 2013. Influence of porosity on compressive and tensile strength of cement mortar. Construction and Building Materials, 40, pp.869-874.

[7] Yu, T.Y., Ing, D.S. and Choo, C.S., 2017. The effect of different curing methods on the compressive strength of eggshell concrete. Indian Journal of Science and Technology, 10(6), pp.1-4.

[8] Kou, S.C., Poon, C.S. and Etxeberria, M., 2014. Residue strength, water absorption and pore size distributions of recycled aggregate concrete after exposure to elevated temperatures. Cement and Concrete Composites, 53, pp.73-82.

[9] A.Buasri, N.Chaiyut, V.Loryuenyong, C.Rodklum, T.Chaikwan, N.Kumphan, K.Jadee,P.Klinklom, W.Wittayarounayut: Sci. Asia, Vol. 38(2012), p. 283-288.

[10] L. M. Correia, R. M. A. Saboya, N. l. de Sousa Campelo, J. A. Cecilia, E. RodrÃ- ezCastellÃ³n, C. l. L. CavalcanteJr, and R. S. Vieira, Bioresource Technology, Vol.151(2014), p. 207-213.

Case Studies of High-strength Lightweight Concrete Using Expanded Siliceous Aggregate

A S Inozemtcev

National Research Moscow State University Of Civil Engineering, Yaroslavskoe shosse, 26, Moscow, 129337, Russia

E-mail: InozemcevAS@mgsu.ru

Abstract. Structural lightweight concrete has various applications including the construction of multi-story buildings, bridges, offshore oil platforms and other reinforced concrete structural elements. Structural lightweight concrete also was used in offshore construction. The development of compositions of high-strength lightweight concrete with an average density of 1100-1500 kg/m^3 may be in demand for the manufacture of most of the typical concrete products with an ordinary strength 20-35 MPa. This study have shown that granular expanded siliceous aggregates can be used as an alternative to hollow microspheres to obtain high mechanical characteristics of lightweight concrete. Based on the principles of producing high-strength lightweight concrete using the example of "Thermogran" granules the possibility of achieving strength of lightweight concrete of more than 45 MPa with an average density of less than 1600 kg/m^3 is shown. In the same time, specific consumption of cement on 1 MPa of concrete strength is 8.1...8.8 kg/MPa.

1. Introduction

Structural lightweight concrete has various applications including the construction of multi-story buildings, bridges, offshore oil platforms and other reinforced concrete structural elements [1, 2]. The use of structural lightweight concrete allows to solve the problems of a large mass of buildings and structures and also stability of construction during operation. Such lightweight concrete with compressive strength in range of 35…41 MPa has been used successfully for almost four decades by precast and prestressed concrete manufacturers in North America [3]. For example, concrete with a density of 1850 kg/m^3 was used to make the floors of the parking complex with 15...20 meter spans in Chicago. The first skyscraper, which was built in 1955 entirely using structural lightweight concrete, is the 18-story building of Dallas Statler-Hilton [4]. Subsequently, such structures became the highest objects in different years. These were the 180-meter towers of the 60-story Marina City in Chicago (1962), the Park Regis Tower of the Australia Square complex 184 meters high (1967); 196-meter Lake Point Tower in Chicago. There are examples of increasing the number of floors of buildings through the use of lightweight concrete [4]. Southwestern Bell Telephone Company has implemented a project to increase the height of an existing building from 14 to 28 floors. Another example is an office building in Ottawa with 22 floors according to the design. Replacing heavy concrete in structures allowed to increase the number of floors to 25. And the height of the Magnolia State Council in Jackson, Mississippi, builders were able to increase from 2 to 6 floors.

During World War II, lightweight concrete was used in offshore construction for the construction of 24 ocean ships and 80 sea barges. [5]. And at the end of the 20th century, such concrete structures were recognized as an effective material for floating oil platforms.

More than 800 bridge structures were constructed using lightweight concrete in North America, such as the 1965 Sebastian Inlet Bridge. [3]. The Chesapeake Bay Bridges are examples of the use of typical lightweight concrete solutions for bridge construction. The first of the two structures was erected in 1953, and the second in 1975.

The relevance of the development of structural or high-strength lightweight concrete is confirmed by research of reducing average density of materials and at the same time increasing strength. The authors [6-9] proposed the use of hollow ceramic or glass microspheres to obtain concretes and mortars. This concrete has a compressive strength of 60...70 MPa and an average density of 1300…1500 kg/m^3 [6, 7]. However, the limited hollow filler is a deterrent to the development of such technology. Such concretes have high techno-economic efficiency [10, 11], but their use is limited to individual construction projects or construction of special-purpose facilities. Existing practice in Russia and the regulatory system for reinforced concrete products [12] show the need for lower strength concrete. An analysis of the Russian regulatory documents for prefabricated and monolithic structures shows that strengths of 20-35 MPa are the most common for heavy concrete. According to the classification of Russian standard GOST 25820-2014 "Light concrete. Technical conditions" lightweight concretes with average density 1100…2000 kg/m^3 and specific strength 10.0-31.8 MPa (table 1) are called structural.

Table 1. Calculation of the minimum required strength of high-strength lightweight concrete for typical building products from reinforced concrete.

№	According to standard		Calculated value	
	Density grade	Strength grade	Specific strength, MPa	Required strength[a], MPa
1	D1100	B12.5	13.6	27.5
2	D1200-D1300	B12.5-B20	11.5-30. 8	30.0-32.5
3	D1400-D1500	B12.5-B30	10.0-28.6	35.0-37.5
4	D1600-D1700	B15-B40	11.8-31.3	40.0-42.5
5	D1800	B20-B40	13.9-27.8	45.0
6	D1900-D2000	B25-B40	17.5-26.3	47.5-50.0

[a] Required strength to provide a specific strength more than 25 MPa.

Thus, we can conclude that structural high-strength lightweight[1] concrete with an average density of more than 1100 kg/m^3 can be used for the manufacture of typical reinforced concrete. That is, the development of compositions of high-strength lightweight concrete with an average density of 1100-1500 kg/m^3 may be in demand for the manufacture of most of the typical concrete products with an ordinary strength 20-35 MPa. And concrete compositions of the D1600-D2000 brand with a strength of 40-50 MPa are required for special structures and construction in earthquake-prone areas. Based on this, the practical task in the technology of high-strength lightweight concrete is to achieve a minimum average density with sufficient strength characteristics. The solution to this problem is to use aggregates alternative to hollow microspheres.

2. Material and methods

The extensive mineral resources base of Russian regions forms the prerequisites for the development of technologies for the production of expanded aggregates based on local raw materials. One of the promising theme for the construction industry is the manufacture of granular glass-ceramic aggregates from glass or amorphous volcanic rocks. [13, 14]. Expanded siliceous aggregate granules have the following properties [15]:

[1] High-strength lightweight concrete is the concrete with specific strength more than 25 MPa.

- Bulk density – 140…180 kg/m3;
- Compressive strength – 0,7…7,0 MPa;
- Water absorption – less than 2%;
- Fractional composition – 0.2…15.0 mm.

Granular siliceous lightweight sand "Thermogran" (EGA) of fraction 0...1 mm was used in this study (figure 1).

Figure 1. Granular expanded siliceous aggregate.

Figure 2. Servo-hydraulic press «Advantest 9».

Portland-cement CEM-I 42.5 (PC), quartz sand fraction of 0.16-0.63 mm (QS), stone powder (SP) with a specific surface area of 700-800 m^2/kg, microsilica (MS), polycarboxylate plasticizer «Melflux 1641F» (A) were used for preparation of concrete mixture. The mixing of concrete was carried out using a mortar mixer «Automix». The servo-hydraulic press «Advantest 9» (figure 2) was used for determination of strength in 1 day after heat-humid treatment according to ASTM C39/C39M-14. The structure of HSLWC was researched using a microscope «Eclipse MA200».

3. Results and discussion

In this work, high-strength lightweight concrete compositions were obtained by partial replacement of aggregate (silica sand) in heavy fine-grained concrete with granular lightweight aggregate with a ratio of components according to table 2.

Table 2. Volumetric content of components for structural lightweight concrete.

№	Composition	Content						
		PC	MS	SP	EGA	QS	W	A
1	Com. 1	0,200	0,040	0,010	0,527	0,017	0,219	0,002
2	Com. 2	0,192	0,037	0,009	0,507	0,015	0,238	
3	Com. 3	0,129	0,036	0,075	0,496	0,045	0,217	
4	Com. 4				0,541	–		

The studied compositions of structural lightweight concrete have a high degree of filling of lightweight fraction (0.541), which allows a low average density. The introduction of expanded siliceous aggregate in the indicated amount into a finely dispersed cement-mineral matrix forms a close-packed material structure (Figure 3) with a saturated gas phase. Features of the technology of producing expanded siliceous aggregate provide the formation melted shell on the surface of the granules. This helps to achieve low water absorption and high resistance to external loads. It also reduces the water demand of mixtures and allows to not saturate the granules with water to ensure mobility observing W/C-ratio (0.35...0.54).

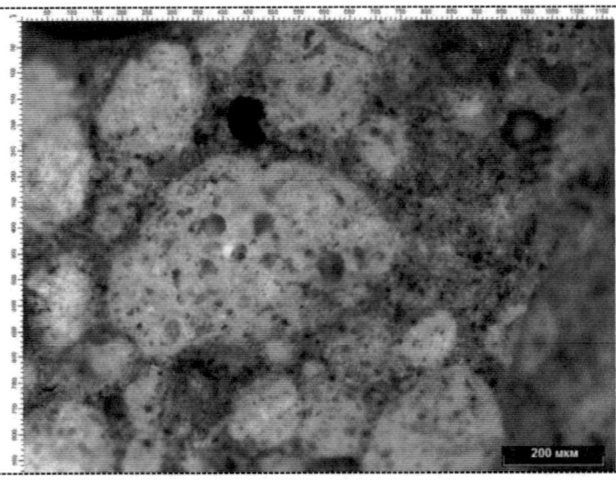

Figure 3. Structure of lightweight concrete on expanded granular aggregate (magnification ×50).

It is established that the filling of cement high-strength matrix by expanded siliceous aggregate provides the formation of a structure with high strength properties (Figure 4). The use of expanded siliceous aggregate allows to reduce the consumption of binder (Figure 5) in contrast to compositions with fine micro-sized aggregate like hollow microspheres [16]. The compressive strength of high-strength lightweight concrete on expanded granular siliceous aggregate is 48.4...52.7 MPa with an average density of 1420...1620 kg/m^3.

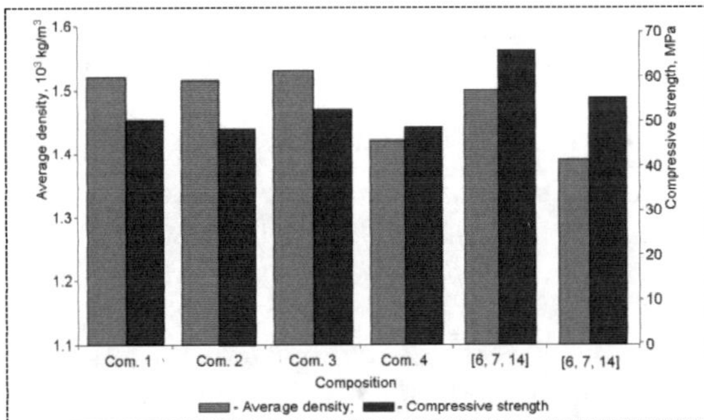

Figure 4. Physico-mechanical properties of concrete.

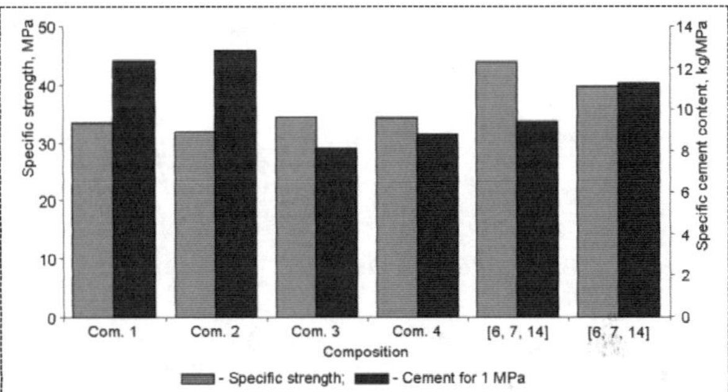

Figure 5. Specific characteristics of concrete.

The calculation of the specific indicators of the studied compositions of structural lightweight concrete (Figure 5) shows the possibility of achieving of specific strength of more than 25 MPa. In this case, the most important indicator of the specific consumption of cement, which shows the consumption of binder on 1 MPa of concrete strength, can be reduced from 12.4…12.9 to 8.1...8.8 kg/MPa.

Thus, the study of lightweight concrete on expanded granular siliceous aggregate shows the possibility of obtaining composites with high physical and mechanical properties. This concrete has a lower consumption of cement and a lower specific consumption of binder per unit of strength compared to high-strength lightweight concrete on hollow microspheres. Based on this, such aggregate prepared from local raw materials is a good alternative to expanded ceramic aggregates and hollow microspheres for structural lightweight concrete due to the lower cost of production.

4. Conclusion

Studies have shown that granular expanded siliceous aggregates can be used as an alternative to hollow microspheres to obtain high mechanical characteristics of lightweight concrete. Based on the principles of producing high-strength lightweight concrete using the example of "Thermogran" granules the possibility of achieving strength of lightweight concrete of more than 45 MPa with an average density of less than 1600 kg/m^3 is shown. In the same time, specific consumption of cement on 1 MPa of concrete strength is 8.1...8.8 kg/MPa.

5. References

[1] Kenneth S, Harmon P E 2006 *Engineering properties of structural lightweight concrete* (Carolinastalite company – united states)
[2] Faust T., Konig G. 1998 High strength lightweight aggregate concrete *Proceedings 2nd International Ph.D. Symposium in Civil Engineering* (Budapest: Technical University of Budapest) p 8
[3] Holm T A, Bremner T W 2000 *State-of-the-Art Report on High-Strength, High-Durability Structural Low-Density Concrete for Applications in Severe Marine Environments*
[4] Zareef M A M E 2010 *Dissertation. Conceptual And Structural Design Of Buildings Made Of Lightweight And Infra-Lightweight Concrete* (Berlin) p 119
[5] Holm T A, Bremner T W , Vaysburd A 1988 Carbonation of marine structural lightweight concrete *Second International Conference on Performance of Concrete in Marine Environment* Andrews St N B (Canada. ACI SP) p 109
[6] Inozemtcev A S 2015 High-strength lightweight concrete mixtures based on hollow microspheres: technological features and industrial experience of preparation *IOP Conference Series: Materials Science and Engineering* **71(1)** 012028
[7] Inozemtsev A S, Korolev E V 2017 Features of the defectiveness of nanomodified high-strength lightweight concrete based on hollow microspheres *Key Engineering Materials* **743** 68–72
[8] Oreshkin D V, Semenov V S, Rozovskaya T A 2014 Light-weight backfill mortars with antifreeze additives for the permafrost conditions *Neftyanoe khozyaystvo - Oil Industry* **4** 42–45
[9] Perfilov V A, Oreshkin D V, Semenov V S 2016 Environmentally Safe Mortar and Grouting Solutions with Hollow Glass Microspheres *Procedia Engineering* **150** 1479–1484
[10] Inozemtcev A S, Korolev E V 2014 Technical and economical efficiency for application of nanomodified high-strength lightweight concretes *Advanced Materials Research* **1040** 176–182
[11] Inozemtcev A, Duong T Q 2019 Technical and economic efficiency of materials using 3D-printing in construction on the example of high-strength lightweight fiber-reinforced concrete *E3S Web of Conferences* **97** 02010
[12] Reference and legal system "Guarantor": [Electronic resource] NPP "Garant-Service". [updated 2019 Oct 24; cited 2097 Dec 11]
[13] Orlov A D 2015 Optimized single-stage technology of granular foam glass based on low-temperature synthesis of glass phase *Building materials* **1** 24–26
[14] Orlov A D 2014 Foam glass ceramic from mineral raw materials: the new one-stage technology "Thermogran" based on low-temperature synthesis of glass phase and its prospects *Bulletin of SIC Construction* **11** 40–45
[15] Orlov A D , Nezhikov A V 2017 Foam glass ceramics as a filler for high-tech lightweight concrete *Bulletin of SIC Construction* **14** 163–171
[16] Inozemtcev A S 2014 Average density and porosity of high-strength lightweight concrete *Magazine of Civil Engineering* **51(7)** 31–37

Acknowledgments
This work was carried out within the framework of President of the Russian Federation grant for state support of young Russian scientists - candidates of sciences MK-5950.2015.8.

AUTHOR INDEX

Agrawal, Alok ... 59, 83
Arellano, L G ... 19
Arellano, L. ... 46
Barabanenkov, M Yu ... 40
Baños, A Trejo ... 19
Bermeo, R ... 46
Blackledge, Jonathan ... 11
Boretti, Alberto ... 11, 51
Brcic, M ... 66
Brnic, J ... 66
Calvino, M. ... 46
Castelletto, Stefania ... 11, 51
Chandrakar, Saurabh ... 59
Charerntanom, W ... 72
Cruz-Irisson, M ... 19, 46
Ge, K ... 35
Gupta, Gaurav ... 83
Hasook, A ... 72
Inozemtcev, A S ... 100
Jinnavat, R ... 72
Johnson, Brett C ... 51
Ken, K ... 77
Korolev, D S ... 40
Krscanski, S ... 66
Makarov, M E ... 40
Makhanov, S. S. ... 24
Makhanov, Stanislav S ... 29
Maneewattanapinyo, Pattwat ... 1
Minnullin, R T ... 40
Miranda, A ... 19, 46
Monton, Chaowalit ... 1
Nan, Feng ... 6
Prasoetsopha, N ... 72
Pérez, L A ... 19
Rosa, Lorenzo ... 11
Salazar, F ... 19, 46
Sapegin, A A ... 40
Satoshi, F ... 77
Sharma, Abhishek ... 59, 83
Shigeyuki, D ... 77
Singsang, W ... 72
Subpa-Asa, P ... 77
Suksaeree, Jirapornchai ... 1
Tan, Xuxiang ... 6
Thainoi, P ... 72
Treelop, Apinya ... 1
Trejo, A ... 46
Tuan, Luu Hai ... 29
Vasudevan, Gunalaan ... 93
Veeravatanayothin, Pitsacha ... 1
Wei, Seah Chan ... 93
Williams, W ... 35